Advances in Anatomy, Embryology and Cell Biology publishes critical reviews and state-of-the-art surveys on all aspects of anatomy and of developmental, cellular and molecular biology, with a special emphasis on biomedical and translational topics.

The series publishes volumes in two different formats:

• Contributed volumes, each collecting 5 to 15 focused reviews written by leading experts
• Single-authored or multi-authored monographs, providing a comprehensive overview of their topic of research

Manuscripts should be addressed to
Co-ordinating Editor

Prof. Dr. H.-W. KORF , Zentrum der Morphologie, Universität Frankfurt, Theodor-Stern Kai 7, 60595 Frankfurt/Main, Germany
e-mail: korf@em.uni-frankfurt.de

Editors

Prof. Dr. T.M. BÖCKERS, Institut für Anatomie und Zellbiologie, Universität Ulm, Ulm, Germany
e-mail: tobias.boeckers@uni-ulm.de

Prof. Dr. F. CLASCÁ, Department of Anatomy, Histology and Neurobiology
Universidad Autónoma de Madrid, Ave. Arzobispo Morcillo s/n, 28029 Madrid, Spain
e-mail: francisco.clasca@uam.es

Dr. Z. KMIEC, Department of Histology and Immunology, Medical University of Gdansk, Debinki 1, 80-211 Gdansk, Poland
e-mail: zkmiec@amg.gda.pl

Prof. Dr. B. SINGH, Western College of Veterinary Medicine, University of Saskatchewan, Saskatoon, SK, Canada
e-mail: baljit.singh@usask.ca

Prof. Dr. P. SUTOVSKY, S141 Animal Science Research Center, University of Missouri, Columbia, MO, USA
e-mail: sutovskyP@missouri.edu

Prof. Dr. J.-P. TIMMERMANS, Department of Veterinary Sciences, University of Antwerpen, Groenenborgerlaan 171, 2020 Antwerpen, Belgium
e-mail: jean-pierre.timmermans@ua.ac.be

218
Advances in Anatomy, Embryology and Cell Biology

Co-ordinating Editor

H.-W. Korf, Frankfurt

Series Editors

T.M. Böckers • F. Clascá • Z. Kmiec
B. Singh • P. Sutovsky • J.-P. Timmermans

More information about this series at
http://www.springer.com/series/102

Piotr Dziegiel • Bartosz Pula •
Christopher Kobierzycki •
Mariusz Stasiolek •
Marzenna Podhorska-Okolow

Metallothioneins in Normal and Cancer Cells

 Springer

Piotr Dziegiel
Department of Histology and Embryology
Wroclaw Medical University
Wroclaw, Poland

Department of Physiotherapy
Wroclaw University School
 of Physical Education
Wroclaw, Poland

Christopher Kobierzycki
Department of Histology and Embryology
Wroclaw Medical University
Wroclaw, Poland

Department of Physiotherapy
Wroclaw University School
 of Physical Education
Wroclaw, Poland

Marzenna Podhorska-Okolow
Department of Histology and Embryology
Wroclaw Medical University
Wroclaw, Poland

Bartosz Pula
Department of Histology and Embryology
Wroclaw Medical University
Wroclaw, Poland

Mariusz Stasiolek
Department of Neurology
Polish Mother's Memorial Hospital –
 Research Institute
Lodz, Poland

ISSN 0301-5556   ISSN 2192-7065   (electronic)
Advances in Anatomy, Embryology and Cell Biology
ISBN 978-3-319-27471-3   ISBN 978-3-319-27472-0   (eBook)
DOI 10.1007/978-3-319-27472-0

Library of Congress Control Number: 2016931386

Springer Cham Heidelberg New York Dordrecht London

Abstract

Metallothioneins (MTs) are low molecular weight proteins, which are present in almost all types of organisms. In mammals, four main MT isoforms designated from MT-1 to MT-4 were identified. Their biological role, according to their characteristic structure, was shown to be mostly associated with cellular metabolism of metal ions, especially zinc. Moreover, the available evidence suggests broad regulatory properties of MTs in the control of cell senescence and various pathological processes including neurodegeneration, cardiovascular pathology, metabolic disorders, and various malignancies. This extensive review provides general information on the structure of MT family members and the cellular functions of MT-1, MT-2, and MT-4 isoforms as well as insights into divergent biological roles of MT-3. Due to the involvement of MT molecules in various processes related to carcinogenesis, an organ-specific presentation of current data concerning their potential impact on the progression of various tumors is given. The regulatory role of MT family members in the function of the immune system is also discussed in depth.

Abbreviations

AC	Adenocarcinoma
AD	Alzheimer's disease
ALL	Acute lymphoid leukemia
ALS	Amyotrophic lateral sclerosis
AML	Acute myeloid leukemia
AP-1	Activator protein 1
apo-MT	Apo-metallothionein
ARE	Antioxidant response element
ATM	Ataxia telangiectasia mutated
BAK1	BCL2-antagonist/killer 1
BCC	Basal cell carcinoma
BCNU	1,3-bis(2-chloroethyl)-1-nitrosourea
CAF	Cancer-associated fibroblast
CCT-LC	Chronic cadmium-treated lung cells
CDKN2C	Cyclin-dependent kinase inhibitor 2 C
C/EBP alpha	CCAAT/enhancer binding protein alpha
CGH	Comparative genomic hybridization
CHL	Classical Hodgkin lymphoma
CIA	Collagen-induced arthritis
CIN	Cervical intraepithelial neoplasia
CJD	Creutzfeldt–Jakob disease
CML	Chronic myelocytic leukemia
CNS	Central nervous system
Con A	Concanavalin A
COX-2	Cyclooxygenase-2
CRT	Chemoradiotherapy
DC	Dendritic cell
DCIS	Ductal breast carcinoma in situ
DLBCL	Diffuse large B-cell lymphoma
DMBA	12-Dimethylbenz[a]anthracene
DS	Down syndrome

EAC	Esophageal adenocarcinoma
EAE	Experimental autoimmune encephalomyelitis
ECRG2	Esophageal cancer related gene 2
EMT	Epithelial-to-mesenchymal transition
ESCC	Esophageal squamous cell carcinoma
ET1	Endothelin receptor-1
FIGO	International Federation of Gynecology and Obstetrics
FTC	Follicular thyroid carcinoma
5-FU	5-Fluorouracil
GIF	Growth inhibitory factor
GIST	Gastrointestinal stromal tumor
GSK-3	Glycogen synthase kinase-3
GST-Pi	Glutathione-S-transferase pi
GSTM1	Glutathione-S-transferase M1
GRE	Glucocorticoid-responsive element
HL	Hodgkin lymphoma
HOSE	Human ovarian surface epithelium
HSG	Human salivary gland
HUVECs	Human umbilical vein endothelial cells
IHC	Immunohistochemistry
IDC	Invasive ductal breast carcinoma
ILC	Invasive lobular breast carcinoma
iNOS	Inducible nitric oxide synthase
LCIS	Lobular carcinoma in situ
LDHL	Lymphocyte-depleted classical Hodgkin lymphoma
LMS	Leiomyosarcoma
LOH	Loss of heterozygosity
LPS	Lipopolysaccharide
LRCHL	Lymphocyte-rich classical Hodgkin lymphoma
mAb	Monoclonal antibody
MCHL	Mixed cellularity Hodgkin lymphoma
MCM2	Minichromosome maintenance protein-2
MSI-H	High microsatellite instability
MT	Metallothionein
MTC	Medullary thyroid carcinoma
MTF-1	Metal-regulatory transcription factor-1
MPD	Myeloproliferative disease
MRE	Metal response element
MS	Multiple sclerosis
MTL-5	Metallothionein-like 5 protein
NLPHL	Nodular-lymphocyte predominant Hodgkin lymphoma
NLS	Nuclear localization signal
NK	Natural killer
NO	Nitric oxide

NSCLC	Non-small cell lung carcinoma
NSHL	Nodular sclerosis Hodgkin lymphoma
OMF	Osteomyelofibrosis
OSCC	Oral squamous cell carcinoma
OVA	Ovalbumin
OXA	Oxaliplatin
P-170	P-glycoprotein 170
PCNA	Proliferating cell nuclear antigen
PD	Parkinson's disease
PI3K	Phosphatidylinositol 3-kinase
PKA	Protein kinase A
PKC	Protein kinase C
PPG	Propargylglycine
PTC	Papillary thyroid carcinoma
RCC	Renal cell carcinoma
ROG	Repressor of GATA
ROS	Reactive oxygen species
RNS	Reactive nitric species
SCC	Squamous cell carcinoma
SCLC	Small cell lung carcinoma
SNP	Single nucleotide polymorphism
STK17A	Serine/threonine kinase 17A
SQC	Squamous cell carcinoma
TAM	Tumor-associated macrophage
TCC	Transitional cell carcinoma
Topo II	Topoisomerase-II
TPA	12-O-tetradecanoylphorbol-13-acetate
TS	Thymidylate synthase
TSHR	Thyroid-stimulating hormone receptor
TSST-1	Toxic shock syndrome toxin-1
UV-B	Ultraviolet radiation B
Zn-T1	Zinc transporter-1

Contents

Chapter 1
Introduction

1.1 General Overview

Metallothioneins (MTs) are low molecular weight proteins (6–7 kDa), which were detected in almost all types of organisms ranging from prokaryotes to more complex eukaryotic species (Coyle et al. 2002b; Pedersen et al. 2009; Palacios et al. 2011; Vasak and Meloni 2011). In mammals, four main MT isoforms designated from MT-1 to MT-4 were identified (Mididoddi et al. 1996). The MT-1 and MT-2 members of the family were discovered by Margoshes and Vallee in 1957, who first isolated both isoforms from horse renal cortex in search of a cadmium binding protein (Margoshes and Vallee 1957). Then, at the beginning of the 1990s, the MT-3 isoform (also known as GIF—growth inhibitory factor) was extracted from the brain of rats suffering from experimentally induced Alzheimer's disease (Uchida and Tomonaga 1989; Uchida et al. 1991). Following this finding, Quaife et al. discovered the MT-4 isoform in the stratified squamous epithelia of the skin and upper parts of the respiratory tract (Quaife et al. 1994). Most of our knowledge concerning MTs' biology stems from research focusing on the role of MT-1 and MT-2 isoforms (MT-1/2), which are ubiquitously expressed in almost all cells of the body (Davis and Cousins 2000; Krizkova et al. 2009b, 2012; Babula et al. 2012). The functions of MT-3 and MT-4 are currently under intense investigation.

Nevertheless, a substantial involvement of MTs has been implicated in multiple biological processes including regulation of zinc absorption from intestine and control of zinc serum levels, intracellular zinc homeostasis, metal cation chaperoning, intracellular metal influx regulation, storage and sequestration of metal ions, metal donation to different enzymes, interaction with zinc finger and zinc-based transcriptional factors, cell cycle regulation, toxic metal detoxification, as well as prevention of oxidative and radiation cell damage (Babula et al. 2012). Due to their broad regulatory properties, MTs were also suggested to play an important role in cell senescence progression, as well as in various pathological processes including

© Springer International Publishing Switzerland 2016
P. Dziegiel et al., *Metallothioneins in Normal and Cancer Cells*, Advances in Anatomy, Embryology and Cell Biology 218, DOI 10.1007/978-3-319-27472-0_1

neurodegeneration (Hashimoto et al. 2011; Michael et al. 2011; Manso et al. 2012), cardiovascular pathology (Guo et al. 2009; Cong et al. 2013; Hu et al. 2013), metabolic disorders (Apostolova et al. 1997; Chen et al. 2001; Li et al. 2004), various types of malignancy (Dziegiel 2004; Pedersen et al. 2009; Pula et al. 2012; Fic et al. 2013), and the vast majority of immune processes. In the last years, some of the postulated pathophysiological effects of MTs were associated with functional polymorphisms of particular genes of MT isoforms (Raudenska et al. 2014).

The general structure of MT members and cellular functions of MT-1, MT-2, and MT-4 isoforms will be discussed in Chap. 2, whereas insights into divergent biological roles of MT-3 will be reviewed in Chap. 3. Due to the involvement of MT molecules in various processes related to carcinogenesis, an organ-specific presentation of their function and potential impact on progression of various tumors will be summarized in Chap. 4. Lastly, the potential regulatory role of MT members for the immune system will be presented in Chap. 5.

Chapter 2
Metallothioneins: Structure and Functions

2.1 The Structure of Metallothioneins

All metallothioneins (MTs) possess a highly conserved amino acid sequence and present only a few structural changes even when isolated from different animal species. In mammals, a single MT molecule is made up of 61–68 amino acids, depending on the isoform (the MT-1, MT-2, and MT-4 isoforms consist of 61–62 amino acids, whereas the MT-3 isoform comprises 68 amino acids), and the protein sequence is composed of up to 20 cysteine (Cys) residues (Vasak 2005; Vasak and Meloni 2011). Furthermore, in mammals, no aromatic amino acids are found in the MT molecules. Protein sequencing has revealed that the MT molecule is a single polypeptide chain, in which the Cys residues are organized in the sequences Cys-X-Cys, Cys-X-X-Cys, and Cys-Cys, where "X" denotes an amino acid other than Cys (Kojima et al. 1976; Huang and Yoshida 1977). The Cys residues are the metal-binding domains of the MT molecule, in which they are juxtaposed with lysine (Lys) and arginine (Arg) amino acid residues and arranged in two thiol-rich sites designated domains α and β (Fig. 2.1). The two metal-binding domains are separated by a non-cysteine-containing sequence often designated as the spacer or linker (Zangger et al. 2001; Babula et al. 2012). The α-domain consists of amino acids 31–68 and is located on the C-terminal edge, whereas the N-terminal β-domain contains amino acids 1–30 (Zangger et al. 2001; Dziegiel 2004). It has been demonstrated that the α-domain is capable of binding up to four, and the β-domain up to three, bivalent metal ions such as zinc, cadmium, mercury, or lead (Coyle et al. 2002b; Duncan et al. 2006). The part of the protein with no bound metal ions is termed apo-metallothionein (apo-MT) or thionein (Coyle et al. 2002b). Metallothioneins are also capable of reacting with up to 12 univalent metal ions (Palmiter 1998; Coyle et al. 2002b). Zinc ions, which naturally occur in the organism, are regarded as the main binding partner of apo-MT. However, other nonessential metal ions occurring pathologically in the organism—such as lead, copper, cadmium, mercury, platinum, chromate, bismuth, and silver—often possess

© Springer International Publishing Switzerland 2016
P. Dziegiel et al., *Metallothioneins in Normal and Cancer Cells*, Advances in Anatomy, Embryology and Cell Biology 218, DOI 10.1007/978-3-319-27472-0_2

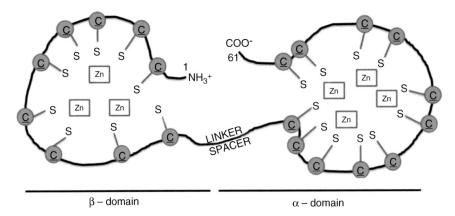

Fig. 2.1 *Metal ion-binding sites of a metallothionein molecule.* The α-domain is capable of binding of up to four, and the β-domain up to three, bivalent metal ions such as zinc, cadmium, mercury, or lead. Adopted and modified according to Nielsen et al. (2007)

higher affinity to the apo-MT-binding sites (Nordberg and Nordberg 2000; Ngu and Stillman 2009; Ngu et al. 2010b; Gumulec et al. 2011; Babula et al. 2012). So far, only iron ions (Fe^{2+}) have been identified to possess lower affinity to the metal-binding sites of the apo-MT domains (Foster and Robinson 2011). Interestingly, only a small proportion of MT molecules was found bound to zinc ions in various organisms. In rat tissues, apo-MT has been shown to constitute up to 54 % of the total amount of MT, whereas higher apo-MT levels were detected in rat cancer cells (Yang et al. 2001). Recent studies have also identified small amounts of sulfide ligands bound to recombinant MT-1 and MT-4 proteins overexpressed in *Escherichia coli* (Capdevila et al. 2005; Tio et al. 2006). Nevertheless, studies analyzing MT proteins in the cytoplasm of mammalian cells have failed to detect sulfide ligands bound to their molecules (Mounicou et al. 2010).

The structure of MT-3 shows a high degree of sequence similarity to other MT molecules (approximately 70 %). However, there are some differences in the sequence that might reflect its functional diversity in comparison to the MT-1 and MT-2 isoforms (Fig. 2.1). In MT-3, a glutamate-rich hexapeptide has been found near the C-terminus containing a Cys-Pro-Cys-Pro motif (amino acids 6–9), which is absent in other MT members (Uchida et al. 1991; Ding et al. 2010). The presence of this fragment, not apparent in other MT members, may result in the reported unique growth-inhibitory activity of the MT-3 molecule and may contribute to other functional differences between the MT-1 and MT-2 isoforms (Uchida and Tomonaga 1989; Uchida et al. 1991; Uchida 1994; Ding et al. 2010; Faller 2010). Taking into account the divergent nature of this isoform, the exact structure and biological functions of MT-3 will be reviewed in Chap. 3.

2.2   Metallothionein Gene Expression and the Regulation of Synthesis

Molecular studies of mouse and human MT genes have permitted the identification and recognition of their expression pattern in various tissues and have allowed the main regulatory mechanisms of its synthesis to be described. In mice, only four MT genes exist localized to chromosome 8 (MT-1, MT-2, MT-3, and MT-4). In humans, however, 17 MT genes have so far been identified in the q13 region of chromosome 16. Of these genes, 13 code for MT-1 and two for MT-2, while two single genes code for each of MT-3 and MT-4 isoforms (Palmiter et al. 1992; Quaife et al. 1994; Mididoddi et al. 1996). Abundant lines of evidence suggest that at least 10 genes encode functional MT proteins: MT-1A, MT-1B, MT-1E, MT-1F, MT-1G, MT-1H, MT-1X, MT-2A, MT-3, and MT-4 (Mididoddi et al. 1996; Werynska et al. 2011) (Fig. 2.2).

Lately, a new functional MT-1M member was identified in the liver (Mao et al. 2012). *MT-1C, MT-1D, MT-1I, MT-1J, MT-1K, MT-1L*, and *MT-2B* are regarded as pseudogenes in humans, as no corresponding proteins have so far been identified. However, their functionality, in most cases, remains unknown (Stennard et al. 1994; Mididoddi et al. 1996). Additionally, a gene called MT-like 5 (MTL-5), which is closely related to the other MT isoforms, has been identified in the testes of mice in the q13 region of chromosome 11 (Olesen et al. 2004). The product of this gene, called tesmin, has been shown to differentially regulate meiosis in male and female cells (Olesen et al. 2004). In summary, the structure and possible MT gene regulatory expression mechanisms are presented in Fig. 2.3.

Metallothionein genes consist of three exons encoding the α-domain (exon 1) and the β-domain (exons 2–3). The regulatory mechanisms of the MT-1 and MT-2 isoforms are the best recognized so far. It has been shown that these isoforms are inducible by several substances and agents, e.g., heavy metals, steroids, cytokines, growth factors, free oxygen, and nitric radicals (Ghoshal et al. 1998, 1999; Jacob et al. 1999; Ghoshal and Jacob 2001; Haq et al. 2003). Indeed, several metal response elements (MRE) (Koizumi et al. 1999; Langmade et al. 2000; Otsuka et al. 2000; Saydam et al. 2002), glucocorticoid-response elements (GRE)

Protein	Aa	1 10	20	30	40	50	60	68
MT-1A	61	MDPNCSCATG	GSCTCTGSCK	CKECKCTSCK	KSCCSCCPMS	CAKCAQGCIC	KGASEKCSCC	A
MT-1B	61	MDPNCSCTTG	GSCACAGSCK	CKECKCTSCK	KCCCSCCPVG	CAKCAQGCVC	KGSSEKCRCC	A
MT-1E	61	MDPNCSCATG	GSCTCAGSCK	CKECKCTSCK	KSCCSCCPVG	CAKCAQGCVC	KGASEKCSCC	A
MT-1F	61	MDPNCSCAAG	VSCTCAGSCK	CKECKCTSCK	KSCCSCCPVG	CSKCAQGCVC	KGASEKCSCC	D
MT-1G	62	MDPNCSCAAA	GVSCTCASSC	KCKECKCTSC	KKSCCSCCPV	GCAKCAQGCI	CKGASEKCSC	CA
MT-1H	61	MDPNCSCEAG	GSCACAGSCK	CKKCKCTSCK	KSCCSCCPLG	CAKCAQGCIC	KGASEKCSCC	A
MT-1X	61	MDPNCSCSPV	GSCACAGSCK	CKECKCTSCK	KSCCSCCPVG	CAKCAQGCIC	KGTSDKCSCC	A
MT-2A	61	MDPNCSCAAG	DSCTCAGSCK	CKECKCTSCK	KSCCSCCPVG	CAKCAQGCIC	KGASDKCSCC	A
MT-3	68	MDPETCPCPS	GGSTCADSC	KCEGCKCTSC	KKSCCSCCPA	ECEKCAKDCV	CKGGEAAEAE	AEKCSCCQ
MT-4	62	MDPRECVCMS	GGICMCGDNC	KCTTCNCKTY	WKSCCPCCPP	GCAKCARGCI	CKGGSDKCSC	CP
		β–domain			α–domain			

Fig. 2.2 *Comprehensive presentation of amino acid sequences of all isoforms of metallothioneins with noticeable division on α- and β-domain. Aa* amino acids. According to UniProt service website

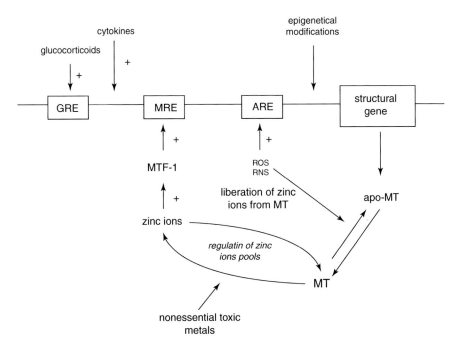

Fig. 2.3 *Possible mechanisms of the regulation of metallothionein gene expression. apo-MT* apo-metallothionein, *ARE* antioxidant response element, *GRE* glucocorticoid-responsive element, *MTF-1* metal-regulatory transcription factor-1, *MRE* metal response element, *ROS* reactive oxygen species, *RNS* reactive nitric species. Detailed description in the text. Adopted and modified according to Sato and Kondoh (2002)

(Hernandez et al. 2000), and antioxidant response elements (ARE) have been noted in the promoter region of the MT-1 and MT-2 genes (Campagne et al. 2000; Bi et al. 2004).

Metal ions, especially zinc ions that occur naturally in the organism, seem to be the most potent inducers of MT-1/2 expression. They have been shown to bind to metal-regulatory transcription factor-1 (MTF-1), which interacts with the DNA via its six zinc finger C2H2 domains to the MRE sequence in the promoter regions of MT-1/2 genes. This binding subsequently results in the initiation of gene transcription (Langmade et al. 2000; Otsuka et al. 2000; Saydam et al. 2002). MRE elements have also been identified in the promoter regions of the MT-3 gene, though contrary results exist concerning possible expression induction of this isoform by metal ions (Heuchel et al. 1994; Chapman et al. 1999; Garrett et al. 2002). MTF-1 is also responsible for the basal expression of MT genes and so far is the only identified mediator of their sensitivity to metal ions (Heuchel et al. 1994; Ghoshal et al. 1999; Ghoshal and Jacob 2001).

Toxicological studies have revealed that metal ions other than zinc may also induce MT-1/2 gene expression. Increased amounts of MT-1/2 were found in the liver, kidney, and intestines of experimental animals following parental or dietary exposure to cadmium, mercury, or zinc (Vasak and Meloni 2011). Although metal ions other than zinc are capable of inducing MT-1/2 expression, this mechanism differs from the induction by zinc ions, discussed above, since they bind directly to MTF-1. The nonessential metal ions cannot activate the MTF-1, but due to their higher affinity to MT-1/2 proteins, they are capable of displacing zinc ions from MT-1/2 molecules and increasing free intracellular zinc levels (Koizumi et al. 1999; Murata et al. 1999). Subsequently, the free zinc ions bind to the MTF-1, leading to activation of MT-1/2 gene transcription (Koizumi et al. 1999; Murata et al. 1999; Lichtlen and Schaffner 2001).

Oxidative stress induced by various factors and conditions has been shown to elevate MT expression, independently of the mechanism involving the MRE (Vasak and Meloni 2011; Babula et al. 2012). The generation of free radicals such as hydrogen peroxide (H_2O_2) by various factors results in the oxidation of the MT molecule and the subsequent release of its bound zinc ions, which ultimately lead to MTF-1 activation (Andrews 2000; Nguyen et al. 2003). It has been shown that catecholamines (Gauthier et al. 2008; Eibl et al. 2010), tissue hypoxia (Murphy et al. 2008; Kojima et al. 2009), physical exercise (Podhorska-Okolow et al. 2006), or hypothermia (Park et al. 2013) may induce MT-1 and MT-2 gene expression. MT-1/2 transcription may be also regulated by glucocorticoids via their direct binding to the GRE in the promoter region of MT genes (Davis and Cousins 2000; Hernandez et al. 2000). High glucose levels have recently been found to induce MT-1 and MT-2 expression in human umbilical vein endothelial cells (HUVECs) on stimulation of endothelin receptor-1 (ET1) (Apostolova et al. 2001). Furthermore, epigenetic regulation of MT-1, MT-2, and MT-3 gene expression via DNA methylation or histone modifications has been observed in the cancer cells of neoplastic diseases such as esophageal, gastric, and prostate cancers (Deng et al. 2003a, b; Smith et al. 2005; Han et al. 2013).

2.3   Localization and Main Functions of MT Proteins

Metallothioneins have been detected in various normal and pathological cells (Dziegiel 2004; Pula et al. 2012; Werynska et al. 2013a), as well as in blood serum (Nordberg et al. 1982; Ghoshal et al. 1998; Adam et al. 2010; Kruseova et al. 2013), where they are capable of regulating and mediating several important cellular processes (Fig. 2.4). Functional MT-1/2 isoforms have mainly been found in cellular cytoplasm and in some organelles, with their expression predominantly noted in mitochondria (Banerjee et al. 1982). Their concentration is strongly dependent on the oxidative status of these organelles. Due to the small molecular size of MTs, they can be transported through the outer membrane of mitochondria and regulate the permeability of their inner membrane (Simpkins et al. 1998). The presence of MT isoforms has also been observed in lysosomes: MT-1 and MT-2 have been shown to protect this organelle from oxidative stress by reducing the iron-catalyzed intralysosomal peroxidative reactions (Baird et al. 2006). Moreover,

MT-3 has been shown to mediate proper lysosome function in astrocytes and neurons by affecting lysosomal function and cell viability (Lee and Koh 2010; Lee et al. 2010). The presence of MT-1/2, as well as of MT-3, has been demonstrated in the nuclei of many cells, e.g., hepatocytes (Fig. 2.5) (Cherian and Apostolova 2000; Werynska et al. 2013a). It has been shown that, under conditions of oxidative stress, the MT-1/2 isoforms are rapidly translocated to the cells' nucleus through nuclear pore complexes. It is further known on the basis of many studies that, once localized in the nucleus, MT-1/2 molecules play an important role

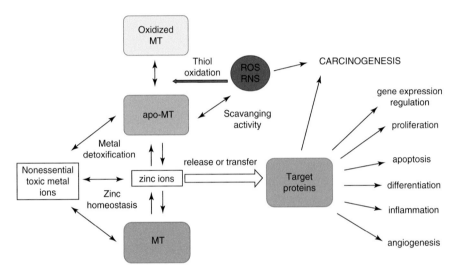

Fig. 2.4 *Cellular processes regulated or mediated by metallothioneins. apo-MT* apo-metallothionein, *ROS* reactive oxygen species, *RNS* reactive nitrogen species

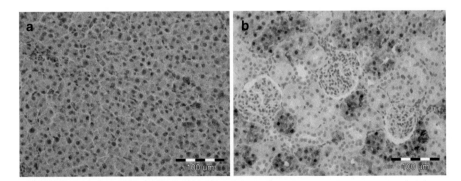

Fig. 2.5 *Immunohistochemical demonstration of metallothionein 1/2 in cytoplasm and nuclei of normal human hepatocytes* (**a**) *and proximal tubule cells in human kidney* (**b**). Staining was performed according to Cherian et al. (2003). Archival sections from the Department of Histology and Embryology, Wroclaw Medical University, Wroclaw, Poland

in cell proliferation and differentiation (Apostolova et al. 2000; Cherian and Apostolova 2000; Chen et al. 2004; Nzengue et al. 2009), as well as in genotoxicity and cell apoptosis (Gunes et al. 1998; Apostolova et al. 1999; Santon et al. 2006). Moreover, nuclear MT-1/2 expression has been noted in hepatocytes (Tsujikawa et al. 1991), myoblasts (Apostolova et al. 2000), and tumor cells of various malignancies (Surowiak et al. 2007; Szelachowska et al. 2008). Similarly, the observed different localizations of MT-3 in the cellular cytoplasm and the nucleus seem to play a significant role in protection against DNA damage and in regulation of transcription (Chen et al. 2002; Werynska et al. 2013a).

Aside from the expression of MTs in various cells, the presence of these proteins has also been detected in blood, with concentrations varying from 0.01–1.0 ng/l in serum (Nordberg et al. 1982) to 0.51–1.86 ng/ml in plasma (Milnerowicz et al. 2009). The assessment of MT levels in patients' blood may become a promising marker for diagnostic and prognostic estimation of therapy efficacy for childhood tumors—in the case of which serum MT levels have been observed to be elevated approximately five times over those of healthy children controls (Krizkova et al. 2010). Sabolic et al. suggest that blood MTs may originate from damaged cells via protein leakage through cell membranes or in the complete demise of cells (Sabolic et al. 2010). However, the functional significance of blood MTs still needs to be determined.

The metal-binding abilities of MTs and their differentiated and vast cellular localizations (due to their small molecular size) give rise to their multifunctionality in various cellular processes (Sabolic et al. 2010; Vasak and Meloni 2011; Babula et al. 2012). Apart from their regulatory and protective roles in numerous normal cells, in some circumstances—such as during carcinogenesis—they may contribute to tumor progression (Dziegiel 2004). MTs are thus often referred to as multipurpose proteins with two faces (Coyle et al. 2002b; McGee et al. 2010). Abundant lines of evidence indicate that these small proteins are involved in several key processes, such as detoxification of heavy metal ions, scavenging of reactive oxygen species (ROS) and of reactive nitric species (RNS), differentiation, proliferation, regulation of cell death, migration, and invasiveness of cancer cells, as well as angiogenesis (Lee and Koh 2010; McGee et al. 2010; Sabolic et al. 2010; Zbinden et al. 2010; Vasak and Meloni 2011). Interestingly, although MTs exert numerous functions, mice lacking MT-1 and MT-2 proteins (MT-null mice), as well as transgenic mice overexpressing these isoforms under normal conditions, show no gross phenotypic or reproductive abnormalities (Sabolic et al. 2010). The significance of MT-1 and MT-2 expression in the abovementioned processes seems to become first apparent under conditions of pathological stress (Takano et al. 2004).

2.4 Detoxification of Metal Ions

MT-1 and MT-2 have been shown to bind various metal ions, univalent as well as bivalent. An ability to detoxify such compounds has thus been ascribed to these proteins (Sabolic 2006; Sabolic et al. 2010). It has been noted that apo-MT molecules may connect with 7–9 bivalent ions (e.g., zinc or cadmium). However, they may also bind up to 12 copper or even 18 mercury ions in experimental conditions (Nielson et al. 1985; Palumaa et al. 2002, 2003, 2005; Meloni et al. 2006). MT molecules may bind different metal ions at the same time (Palumaa et al. 2002; Romero-Isart and Vasak 2002; Palacios et al. 2011). Spontaneous oligomerization via the disulfide bonds of MT molecules' α-domains has been also reported (Zangger et al. 2001). Such aggregates, isolated from horse and rabbit kidney, have been found to be capable of binding more metal ions than could be predicted from their single polypeptide chain structure (Zangger et al. 2001; Wilhelmsen et al. 2002). The exact affinity of particular metal ions has been determined in various studies. However, most toxic ions (such as cadmium, lead, and mercury) have been shown to possess higher affinity toward MT molecules in the majority of studies (Waalkes et al. 1984; Nielson et al. 1985; Sabolic et al. 2010). These metal ions may displace zinc or other lower affinity metal ions bound to MTs, which may lead to alterations in crucial cellular process such as transcription or translation (McGee et al. 2010). The protective effect of MT-1/2 in the majority of circumstances was mainly mediated by the release of zinc ions from MT-1/2 molecules and by the subsequent activation of MTF-1, leading to an increase in the synthesis of new MT proteins. Furthermore, the released zinc ions antagonize the pro-oxidative effect of other toxic metals, such as cadmium, by shifting the redox state toward the reducing/antioxidative effect of the MT-1/2 molecules (Sabolic et al. 2010).

As mentioned above, metal ions are capable of inducing MT expression in many mammalian tissues, including the liver, kidney, testes, and intestine (Sabolic et al. 2010; Babula et al. 2012). Increased expression of MTs has been noted in other organisms. It has been shown that the assessment of MT levels in the gills of various animal species may be a useful and effective biomarker for monitoring environmental contamination with toxic metals (Hamza-Chaffai et al. 1999; Hayes et al. 2004; Smaoui-Damak et al. 2009). Moreover, the protective role of MTs against heavy metal intoxication is additionally supported by their ROS scavenging ability (Chiaverini and De Ley 2010). Abundant experimental data underlie the importance of MT-1 and MT-2 as protective agents against intoxication with nonessential heavy metal ions.

2.4.1   Role of Metallothioneins in Preventing Cell and Organ Toxicity Induced by Cadmium Ions

Hitherto, the relationship between MTs and cadmium, a widely occurring environmental pollutant, has been best studied in relation to cytotoxic effects. Cadmium ions are taken up by animals and humans in contaminated food or inhaled in pollen to the lungs. Upon ingestion, cadmium cations are redistributed from the gastrointestinal tract and transferred to the liver, kidney, and testes. In intoxicated organs, cadmium ions exert their cytotoxic effects by activating ROS generation, subsequently leading to lipid peroxidation, DNA damage, and protein denaturation (Rani et al. 2014). The acute effects of cadmium intoxication, e.g., pulmonary edema, hemorrhage, fulminate hepatitis, and testicular lesions, as well as its chronic effects (nephrotoxicity, osteotoxicity, and immunotoxicity) have been described (Rani et al. 2014).

Experimental data show that pretreatment of animals with a small amount of cadmium increases their resistance to very high doses of cadmium ions, which in normal conditions could induce cell death (Goering and Klaassen 1983). Isolation of hepatic subcellular fractions in these animals 2 h after injection of the lethal cadmium dose revealed a diminished level of cadmium ions in nuclei, mitochondria, and endoplasmic reticulum and increased levels of cadmium ions in cytosol. Cadmium ions in the cytoplasm were bound to MTs, which had been markedly induced by cadmium pretreatment of the animals (Goering and Klaassen 1983). Further studies confirmed that high liver MT levels are capable of diminishing the hepatotoxic effect of cadmium ion administration to both newborn and adult rats (Goering and Klaassen 1984; Mukhopadhyay et al. 2009). Genetically modified MT-1/2 null animals (Liu et al. 1996; Zheng et al. 1996a, b; Habeebu et al. 2000a) and experimental models based on MT overexpression (Liu et al. 1995) confirmed the critical role of MT-1/2 in protecting against the acute and chronic effects of cadmium ions on liver toxicity.

Moreover, increases in MT-1 and MT-2 levels have also been reported in the lung following cadmium treatment via inhalation or intratracheal instillations (Hart et al. 1989, 1995, 2001; Kenaga et al. 1996; Potts et al. 2001). As was shown in the liver, pre-exposure to low doses of cadmium induces MT levels in the lungs of male Lewis rats, resulting in an increased tolerance of these animals to higher toxic doses of this metal ion (Hart et al. 1989). Potts et al. suggested that the most prominent change in cadmium-treated epithelial lung cells is the upregulation of MT expression, which may sequester cadmium ions and diminish the subsequent cytotoxic effects of generated ROS (Hart et al. 2001; Potts et al. 2001). Although the cells adapted to cadmium challenge by increasing MT-1/2 expression, they were also characterized by reduced ability of DNA repair and became more resistant to apoptotic stimuli (Hart et al. 2001; Potts et al. 2001). Human chronic cadmium-treated lung cells (CCT-LC) were shown to acquire a phenotype similar to that of cancer cells and this phenomenon was accompanied by an increase in MT-1/2 expression. This transformation occurs despite the cells' ability to adapt to chronic cadmium exposure (Person et al. 2013). Indeed, the increase in MT-1/2 expression in lung alveolar cells was accompanied by decreased apoptosis of these cells and may therefore promote tumor development (Hart et al. 2001). Similarly, a decrease in E-cadherin expression in human lung cells was reported by Pearson et al. (2003) and was regarded as a significant risk factor for lung cancer development. Chronic cadmium intoxication and induction of MT-1/2 expression could thus contribute to the carcinogenesis process, quite apart from the protective role of these proteins in normal cells (Hart et al. 2001).

Animal models and in vitro studies allow the possible actions of MT-1/2 in protecting the kidneys against cadmium intoxication to be recognized. Cadmium ions may enter the tubular cells of kidneys via the basal and the luminal cell membranes (Zalups and Ahmad 2003). Acute nephrotoxicity has been observed upon *i.v.* injection of the cadmium-MT (Cd-MT) complex in mice (Nordberg et al. 1975). However, in this experimental model, the acute effects of cadmium intoxication were caused by the damaging effect exerted by the cadmium-MT complex upon its uptake from primary urine into proximal tubule cells (Nordberg et al. 1975). Nevertheless, under normal conditions, this mechanism is not observed, as free cadmium ions that are not bound to a complex with MTs are mostly absorbed from the interstitial fluid. Interestingly, it has been shown that MT-null mice, although presenting much lower cadmium accumulation in the kidney, manifest much more rapid and severe impairment in kidney function, as compared to wild-type mice, indicating a protective function of endogenously expressed MTs (Liu et al. 1998). Moreover, it seems that most protective effects can be seen in renal tubular proximal cells, which express higher levels of MT-1/2 as compared to tubular distal cells (Fig. 2.4). In addition, an in vitro study showed that the bound Cd-MT complex was less toxic than $CdCl_2$ to the cultured rat proximal tubule cells, as well as to pig renal proximal tubular cells (LLC-PK1 cell line), contradicting the observations of Nordberg et al. (Liu et al. 1994). The role of cadmium and Cd-MT complex in chronic nephrotoxicity was investigated by administrating equal amounts of cadmium ions in a form of $CdCl_2$ or Cd-MT complex for 10 months to male Wistar rats (Groten et al. 1994). In this experimental setting, animals treated with $CdCl_2$ were characterized by a much higher nephrotoxicity, whereas the Cd-MT group only showed a slight increase in urinary gamma-glutamyl transpeptidase activity (a sensitive indicator of ischemic experimental injury) at the end of the experiment (Groten et al. 1994). It therefore seems that it is not the Cd-MT molecules, but the free intracellular cadmium ions, that are responsible for the observed nephrotoxic effects. The increased MT-1/2 expression in rat proximal tubular cells could also protect the cells from exercise-induced apoptosis by scavenging ROS (Podhorska-Okolow et al. 2006).

MTs have also been shown to protect bone tissue from cadmium-induced toxic effects. As in the case of hepato- and nephrotoxicity, MT knock-out mice were characterized by hypersensitivity to cadmium-induced bone injury, as compared to wild-type controls (Habeebu et al. 2000b; Regunathan et al. 2003). Osteocytes were identified to be the cells most affected by cadmium exposure, as increasing levels of

MT-1/2 in response to this toxic metal have been noted in this cell type (Oda et al. 2001).

2.4.2 Role of Metallothioneins in Preventing Cell and Organ Toxicity Induced by Other Metal Ions

Apart from cadmium ions, MT molecules have been shown to mediate resistance toward several other toxic metal ions, such as arsenic, lead, mercury, copper, and chromate. However, in comparison to cadmium ions, the protection provided by MTs against these toxic ions has been far less closely studied. Lines of evidence suggest that MTs are involved in protection against arsenic, which is responsible for numerous toxic and carcinogenic effects (Schuhmacher-Wolz et al. 2009). Inorganic arsenic can be transformed into more toxic methylated arsenicals, which are potent carcinogens causing tumors of the skin, lungs, and urinary bladder (Schuhmacher-Wolz et al. 2009). Arsenic compounds are capable of binding to MT's thiol groups at experimental low pH values of 3.5, as well as at physiological pH 7.0. This indicates that MT-1/2 may detoxify arsenic metal ions (Ngu and Stillman 2006; Ngu et al. 2010a). Moreover, it has been shown that arsenate induces MT-1/2 synthesis (Kreppel et al. 1993). MT-1/2 molecules have been shown to protect kidneys (Liu et al. 2000b), liver, and lungs (Jia et al. 2004a) against arsenic compounds in experimental conditions. In both studies, the severity of the lesions in these organs was significantly higher in the MT knock-out mice compared to the control animals (Liu et al. 2000b; Jia et al. 2004a).

Increases in MT-1 expression have also been noted in response to lead challenge, although a dual effect on expression of this isoform in mice was noted (Yu et al. 2009). An enhancement of MT-1 gene transcription was observed in the liver and kidney; however, MT synthesis was suppressed in the kidneys of the experimental animals (Yu et al. 2009). MT knock-out mice were also more susceptible to lead-induced toxicity, as exposure to this metal significantly impaired renal function in comparison to wild-type mice (Qu et al. 2002). Moreover, MT knock-out mice accumulated less lead in kidneys than did wild-type mice and did not form lead inclusion bodies (Qu et al. 2002). Consistent with the protective role of MTs in lead toxicity, Tokar et al. showed that MT knock-out mice are more sensitive to early-life lead exposure with regard to the frequent generation of testes tumors and of renal and urinary bladder preneoplastic lesions (Tokar et al. 2010).

MT-1 and MT-2 have also been shown to act as protective agents against mercury, a potent neurotoxin (Monnet-Tschudi et al. 2006). Mercury ions possess a high affinity to the thiol groups of various proteins, leading to their inactivation. Thus, the observed induction of MT-1/2 synthesis following mercury intoxication seems to counteract its toxic effects (Chan et al. 1992). So far, MT-1 and MT-2 have been shown to reduce the toxic effects of copper intoxication in the liver, central nervous system, kidneys, and skin (Liu et al. 2000a; Aschner et al. 2006; Brandao et al. 2006; Peixoto et al. 2007; Hwang et al. 2013).

MT-1 and MT-2 have also been shown to sequester copper ions in two hereditary diseases affecting its metabolism, namely, Wilson's disease and Menke's disease (Nartey et al. 1987b; Suzuki-Kurasaki et al. 1997; Klein et al. 1998). Wilson's disease is a rare inherited syndrome presenting with altered copper metabolism characterized by copper deposition in the liver, brain, and cornea. Analysis of the liver of a patient affected by Wilson's disease revealed that copper ions were bound to MT-1/2 molecules (Nartey et al. 1987b). Experimental data from Long–Evans Cinnamon (LEC) rats, which bear a mutation similar to that noted in Wilson's disease, confirmed that MT-1 and MT-2 bind the majority of free copper ions in the lysosomes, where they are transformed over time, leading to subsequent hepatocyte necrosis via oxidative damage (Klein et al. 1998). This knowledge allowed a preventive therapy to be designed, based on chronic administration of zinc, which acts indirectly as a competitive inhibitor of copper ions' entry into the bloodstream. In this therapy setting, zinc ions increase the enterocyte pool of MT-1/2 molecules, which intercept the excess of absorbed copper ions. This prevents their entry into the bloodstream and the subsequent damage of liver and central nervous system. Chronic zinc therapy has replaced penicillamine (a potent metal-ion chelator) as the first-line therapy option in Wilson's disease (Hoogenraad 2006).

MT-1 and MT-2 have also been shown to bind chromate ions. Unlike other nonessential metals, the apo-MT molecule possesses a much higher binding affinity to this potent carcinogen and forms a stable complex once bound (Krepkiy et al. 2003). The mode of action of chromate ions in cells may involve generation of ROS and disruption of protein–DNA interactions (Borthiry et al. 2007, 2008). Chromium (Cr^{6+}) may thus also inhibit MT synthesis by interfering with the formation of the complex of MTF-1 and histone acetyltransferase p300/CBP, which is crucial for MT-1/2 transcription initiation. The decreased cellular levels of MT-1/2 may result in potentiation of the carcinogenic effects of Cr^{6+} due to the decreased antioxidant potential of the cell (Krizkova et al. 2012).

The high affinity of MT-1/2 to metal ions may also be the reason for the inactivation of alkylating drugs whose cytotoxic effect depends on the presence of heavy metal compounds (e.g., cisplatin, carboplatin) (Andrews et al. 1987; Shimoda et al. 2003; Choi et al. 2004). This effect may contribute to chemotherapy failure in some cancer types (Surowiak et al. 2003, 2007).

2.5 The Role of Metallothioneins Under the Conditions of Oxidative Stress

Abundant lines of evidence point to the role of MT-1/2 in diminishing the effects of oxidative stress, due to the action of free radicals such as ROS or RNS (Valko et al. 2006). These shortly lived molecules are characterized by the presence of at least one unpaired electron and may be generated upon action of different physical factors (UV, gamma or X-ray radiation, chemical reactions catalyzed by metals) or during various biological processes (inflammatory reactions, mitochondrial respiration). Free radicals may be beneficial in inflammatory and immune reactions, but, produced in excess, they may lead to damage of different cellular structures, ultimately leading to cell death or neoplastic transformation (Valko et al. 2006). It was demonstrated that their damaging effects may be counterbalanced by antioxidant molecules such as MT-1 and MT-2 isoforms (Krizkova et al. 2009a, 2012; Chiaverini and De Ley 2010).

The thiolate cluster of MT-1/2 proteins is responsible for these molecules' redox potential, and is dependent on the stability of the zinc/thiolate binding, which in turn modulates the mobility of zinc ions and their transfer to other zinc-dependent proteins (Maret and Vallee 1998). The observed induction of MT-1/2 expression by free radicals has led to the suggestion that these proteins may protect cells from oxidative stress (Andrews 2000; Nguyen et al. 2003). The antioxidant properties of MT-1/2 have been confirmed in numerous in vitro and in vivo studies. Thornalley and Vasak observed for the first time that, in rabbit liver, MT-1 scavenged free hydroxyl and superoxide radicals much more effectively than bovine serum albumin, which was used as a control in this cell-free experimental setting (Thornalley and Vasak 1985). The antioxidative properties enable MT-1/2 to decrease the DNA damage caused by hydroxyl radicals (Abel and de Ruiter 1989). Furthermore, induction of MT-1/2 expression by $ZnCl_2$ pretreatment of HL-60 human promyelocytic leukemia cells and V79 Chinese hamster cells confirmed the ROS-scavenging ability of MT-1/2 (Chubatsu et al. 1992; Quesada et al. 1996). In both cell types, elevated MT-1/2 levels abrogated oxidative stress and reduced the DNA damage in comparison to the control cells (Chubatsu et al. 1992; Quesada et al. 1996). MT-1/2 also protected mouse embryonic fibroblasts (NIH 3T3) from tert-butyl hydroperoxide toxicity—a potent inducer of free radicals (Schwarz et al. 1995). Moreover, MT-2A, overexpressed in lymphocytes, has been shown to protect them from UV radiation-induced damage (Yang et al. 2007).

The ability of MT molecules to transfer from cytoplasm to nucleus has been found to be strongly dependent on the redox state of the nucleus (Apostolova et al. 2000; Ogra and Suzuki 2000). Since the cell cycle may be affected by redox status, it has been suggested that MT-1/2 molecules may protect the cell's nucleus from excess free radicals and thus regulate and warrant the progress of the cell cycle (Takahashi et al. 2005).

Interestingly, although in vitro studies point to the antioxidant role of MT-1/2, in vivo experiments have not confirmed the results of the in vitro experiments. The use of MT knock-out mice permitted an analysis of the role of MT isoforms under in vivo conditions. The levels of antioxidant proteins and molecules (superoxide dismutase, catalase, or glutathione peroxidase and glutathione) did not differ in MT-null mice exposed to oxidative stress induced by gamma-irradiation or 2-nitropropane, compared to control wild-type mice (Conrad et al. 2000). The extent of oxidative damage to DNA, lipids, and proteins in the liver of both mouse types was comparable, although the levels of MT-1/2 in the liver of the wild-type mice increased significantly. Furthermore, when mice were exposed to whole-body irradiation, no differences were observed in their survival time, even when the animals were subjected to zinc pretreatment in order to increase MT-1/2 expression levels (Conrad et al. 2000). Also the study of Davis et al., who used MT knock-out and MT-1/2 overexpressing mice also did not produce convincing results regarding the antioxidative role of MT-1/2 in vivo (Davis et al. 2001). Hepatotoxicity, as measured by serum alanine aminotransferase activity, histological analyses, and hepatic thiol levels, was greater in the knock-out mice than in the controls 12 h after carbon tetrachloride treatment. However, no differences were observed at later time points up to 48 h. Hepatotoxicity was also similar between MT-overexpressing and control mice, and the dietary zinc treatment provided no further protection (Davis et al. 2001).

Although the abovementioned studies do not confirm the antioxidant role of MT-1/2, several other in vivo experiments have noted significant differences, suggestive of the protective role of MT expression in some diseases. In a model of acetaminophen-induced liver injury, MT-1/2 prevented the generation of ROS, rather than functioning as a ROS scavenger (Saito et al. 2010). The reactive acetaminophen metabolite, N-acetyl-p-benzoquinone imine, was effectively trapped by MT-1/2 molecules by covalent bindings. This prevented the binding of N-acetyl-p-benzoquinone imine to cellular proteins and thus blocked the further actions that would lead to mitochondrial dysfunction and nuclear DNA damage (Saito et al. 2010). In addition, MT-1 and MT-2 have been shown to protect the cell against DNA damage in liver and bone marrow in mice with high fat-diets (Higashimoto et al. 2009). MT-1 and MT-2 also have protective effects on gastric mucosa during inflammation caused by infection with *Helicobacter pylori* (Mita et al. 2008). MT-1/2 inhibited the activation of cyclooxygenase-2 (COX-2) in an experimental model of collagen-induced arthritis (Youn et al. 2002), and inducible nitric oxide synthase (iNOS) in a model of cryogenic brain cortex injury (Penkowa et al. 2006), which supports the antioxidant role of these MT isoforms. Mice overexpressing the MT-1/2 isoforms were relatively resistant to oxidative stress. However, MT knock-out mice have been shown to be sensitive to oxidative stress-induced carcinogenesis of the skin and liver cells in vivo (Suzuki et al. 2003; Waalkes et al. 2006). Due to the scavenging ability of MT-1/2, these molecules have been shown to mediate cancer cells' resistance to certain chemotherapeutics, such as irinotecan (Chun et al. 2004), adriamycin (Hatcher et al. 1997), and doxorubicin (Yap et al. 2009), as well as radiotherapy (Cai et al. 1999; Smith et al. 2006).

2.6   Metallothioneins and Apoptosis

Apoptosis plays an important role in numerous physiological as well as pathological processes. Under normal conditions, unnecessary cells are eliminated from tissues, allowing normal functioning. Abnormalities and inhibition of this process may lead to the development of many diseases—especially autoimmune or neo-plastic disorders (Khan et al. 2014). Apoptosis has been shown to be modulated by the actions of ROS and zinc ions (Formigari et al. 2007, 2013). This essential metal ion has been shown in eukaryotic cells to regulate the activity of various enzymes and transcription factors, thus influencing several key processes, such as differen-tiation, cell growth, and apoptosis (MacDonald 2000). Zinc ions also function as protective agents for enzymes by inhibiting the oxidation of their sulfhydryl groups (Powell 2000). Due to the free radical scavenging and the metal-binding ability of MT-1/2, these proteins have been shown to strongly influence apoptosis (Coyle et al. 2002b; McGee et al. 2010). Depending on the cell type, MT-1/2 may protect DNA from UV-induced damage as a health-promoting effect (McGee et al. 2010). However, from another point of view, the elevated expression of MT-1/2 in cancer cells has been shown to protect them from chemotherapy and radiotherapy (Andrews et al. 1987; Hishikawa et al. 1997; Cai et al. 1999; Bedrnicek et al. 2005).

One of the major regulators of apoptosis is the p53 protein. In many studies, its expression has been shown to strongly depend on zinc ion concentration and ROS levels (Meplan et al. 2000; Fan and Cherian 2002; Ostrakhovitch and Cherian 2004). Furthermore, zinc is required for the proper functioning of the p53 protein, as it stabilizes the protein structure via its binding to the Cys_3His_1 cluster of the DNA-binding domain of the molecule (Meplan et al. 1999). Additionally, zinc ions have been shown to be essential for p53-mediated transcription regulation (Meplan et al. 2000). Taking into account the fact that MT-1/2 may regulate cell zinc and ROS content, they are also capable of regulating the activity of the p53 protein in several ways.

Direct interaction of apo-MT with the p53 protein has also been documented (Ostrakhovitch et al. 2006; Xia et al. 2009). However, MT (apo-MT bound with zinc ions) did not behave in this way and did not affect the transcriptional activity of p53 (Ostrakhovitch et al. 2006). Further studies identified that the interaction is mediated by the free sulfhydryl groups of apo-MT, which interacted with the zinc ion of the p53 protein. Such binding results in the acquisition of a mutant-like phenotype by p53, resulting in the loss of its DNA-binding abilities and in subse-quent apoptosis inhibition (Xia et al. 2009). Moreover, recombinant MT molecules have been found to modulate the structure of the p53 protein in experimental in vitro conditions (Meplan et al. 2000). Furthermore, high levels of MT-1/2 could hypothetically intercept zinc ions from the p53 molecule, inhibiting its DNA-binding capabilities and suppressing its pro-apoptotic function (Palecek et al. 1999). On the other hand, cells lacking MT-1/2 were characterized by high levels of the p53 protein and demonstrated a high sensitivity toward apoptosis-inducing factors (Kondo et al. 1997). Interestingly, Ostrakhovitch et al. demonstrated that only cells with intact function of the p53 protein are capable of inducing MTs' expression upon treatment with metal ions (Ostrakhovitch and Cherian 2004; Ostrakhovitch et al. 2007). In an earlier study, the authors demon-strated that copper-induced apoptosis in MCF-7 cells was strongly dependent on the activity of the p53 protein, whereas the triple negative MDA-MB-231 cells, lacking functional p53, did not respond to such treatment (Ostrakhovitch and Cherian

2004). These results have been confirmed in an experiment performed on the p53 positive MN1 and parental MCF-7 breast cancer cells, which responded to zinc and copper treatment by increasing MRE activity and MTF-1 expression (Ostrakhovitch et al. 2006). Inactivation of p53 in these cells rendered them unresponsive to copper and zinc treatment, and no increase in MTF-1 expression could be noted. Furthermore, the introduction of functional wild-type p53 into the MDD2 breast cancer cell line with a dominant-negative p53 enhanced the ability of zinc ions to increase MTF-1 gene expression (Ostrakhovitch et al. 2007). These results suggest that, in addition to the presented interactions of MT and p53, an additional regulatory loop exists between these proteins, but further research is necessary to corroborate these findings.

Abundant lines of evidence point to the role of MT-1/2 in mediating the activity of *NF*-κB-related pathways. *NF*-κB is a DNA-binding protein complex found in almost all animal cells and involved in the regulation of numerous processes. Its activation has been observed in stress conditions, where it protects cells by activating antiapoptotic genes and proto-oncogenes, allowing cell survival (Gilmore 2006; Perkins 2007). It has been shown that MT-1/2 interacts with the p50 subunit of *NF*-κB and positively regulates its activity in murine fibroblasts (Butcher et al. 2004), cervical cancer HeLa cells (Kim et al. 2003), and MCF-7 breast cancer cells (Abdel-Mageed and Agrawal 1998). On the other hand, it was found that overexpression of MT-2 impaired *NF*-κB activation and sensitized the V-H4 hamster mutant cell line to mitomycin-C (Papouli et al. 2002). The inhibition of MT-2 in these cells, utilizing antisense nucleotides, restored the chemoresistance of the cell line to this therapeutic agent (Papouli et al. 2002). Similarly, Sakurai et al. demonstrated MT-1/2 to be a negative regulator of *NF*-κB activity in mouse embryonic cells (Sakurai et al. 1999). Taking into account the results of all studies performed so far, it seems that the modulation of *NF*-κB activity by MTs is strongly dependent on the cell type, which points to the involvement of yet undiscovered mechanisms. Nevertheless, the increased activity of *NF*-κB in cancer cells on induction of MT-1/2 expression may contribute to carcinogenesis and disease progression (Abdel-Mageed and Agrawal 1998; Kim et al. 2003).

2.7   Metallothioneins and Cell Proliferation

High levels of MT-1/2 have been found in active and proliferating tissues, whether normal or affected by a specific disease process (Cherian and Apostolova 2000; Cherian et al. 2003; Dziegiel 2004; Krizkova et al. 2009b, 2012). In normal tissues, high MT-1/2 levels have been demonstrated by immunohistochemistry (IHC) in proliferating hair follicles of healthy skin (Karasawa et al. 1991) and in the basal cell layer of the epidermis (Zamirska et al. 2012). MT-1/2 has also been shown to mediate proper wound healing of the skin, which was accompanied by changes of zinc ion levels in proliferating cells as well as in serum (Iwata et al. 1999). Zinc ions have been shown to modulate several processes, as they function as cofactors for many proteins, e.g., matrix metalloproteinases and transcription factors, assuring in this way their correct functioning (Chasapis et al. 2012; Krizkova et al. 2012). The metal-binding properties of MT-1/2 allow them to act as possible zinc ion donors for zinc-dependent enzymes and transcription factors functioning via their zinc finger domains. Proteins of the MT-1/2 family play a crucial role in processes such as replication, transcription, and translation (Ostrakhovitch et al. 2007; Pedersen et al. 2009). It has been shown that MT-1 and MT-2 act as metal chaperones, regulating the availability of zinc ions to the zinc finger domains of transcription factors. In many studies, it has been demonstrated that MT-1/2 regulates in this matter the activity of transcription factor IIIA (Zeng et al. 1991b; Petering et al. 2000; Huang et al. 2004), Gal4 (Maret et al. 1997), transcription factor Sp1 (Zeng et al. 1991a; Rana et al. 2008), tramtrack (Roesijadi et al. 1998), and the estrogen receptor (Cano-Gauci and Sarkar 1996).

The abovementioned observations correspond to MT-1/2 localization changes regarding cell cycle stages. In resting cells (G0 phase), MTs can be detected in the cytoplasm, whereas in cells undergoing division, these proteins have been observed to shift to the nucleus. Such a mechanism has been observed, e.g., in case of regenerating hepatocytes (Cherian and Apostolova 2000; Cherian and Kang 2006) and differentiating myoblasts (Apostolova et al. 2000). Furthermore, the high cytoplasmic expression of MTs has been noted in the late G1 phase and at the G1/S threshold, while the highest MT-1/2 concentration in the cell nucleus has been noted in the S and G2 phases (Cherian and Apostolova 2000; Levadoux-Martin et al. 2001). The data derived from normal cells have been confirmed by in vitro experiments on papillary thyroid carcinoma (KAT5) and anaplastic thyroid carcinoma (ARO) cell lines. Cadmium-induced MT-1/2 expression has been found to be related to alterations in the cell cycle, leading to increases in the proportion of cells in the S and G2-M phase and decreases in the number of cells in the G0/G1 phase (Liu et al. 2007, 2009). In human colonic HT-29 cancer cells, the highest MT-1/2 level has been noted in the G1 phase, confirming their proproliferative role (Nagel and Vallee 1995). Furthermore, the recent study of Lim et al. has revealed that downregulation of the MT-2A gene in MCF-7 cells by the use of siRNA particles inhibits cell growth by inducing cell cycle arrest in the G1-phase (G1-arrest), with a marginal increase in cells in sub-G1-phase (Lim et al. 2009). In addition, the MT-2A-silenced cells were characterized by a higher expression of the ataxia telangiectasia mutated (*ATM*) gene, concomitant with a lower expression of the *cdc25A* gene—suggesting that this MT isoform may mediate the progression from the G1 to the S phase of the cell cycle by regulating the activity of the ATM/Chk2/cdc25A pathway (Lim et al. 2009). The results obtained in in vitro experiments have been confirmed by studies on tumor sections. A positive correlation of MT-1/2 expression with that of Ki-67 antigen or PCNA (proliferating cell nuclear antigen) has been noted in formalin-fixed paraffin-embedded tumor sections of breast cancer (Jin et al. 2002; Gomulkiewicz et al. 2010; Wojnar et al. 2011), non-small cell lung cancers (Werynska et al. 2011), squamous cell cancers of the skin (Zamirska et al. 2012), and soft tissue sarcomas (Dziegiel et al. 2005), supporting their role in the regulation of proliferation.

2.8 The Structure and Role of the MT-4 Isoform

The MT-4 isoform is so far the least studied member of the MT family. The MT-4 gene is localized approximately 20 kb upstream from the 5' end of the MT-3 gene in the human and mouse genomes (Quaife et al. 1994). MT-4 is a 62 amino acid long, single-chain protein, similar in its structure to the members of the MT-1/2 isoforms. However, a glutamate insert is present at position 5, unlike in the MT-1 and MT-2 molecules (Quaife et al. 1994). Similarly to the MT-1 and MT-2 proteins, MT-4 is capable of binding up to seven bivalent metal ions. However, its metal-binding domains seem to be characterized by a higher demetallization resistance by EDTA, in comparison to that of the MT-1 molecule (Cai et al. 2005; Meloni et al. 2006). Interestingly, research conducted on *Escherichia coli* has shown that MT-4 has a higher affinity to monovalent copper ions than to bivalent zinc ions, which may significantly affect its function in cells. In turn, the MT-1 protein preferentially binds zinc ions (Tio et al. 2004). Moreover, MT-4 also possesses lower binding affinity to cadmium ions than MT-1 (Tio et al. 2004). Therefore, the MT-4 molecule has also been suggested to be a "copper thionein" in comparison to other family members, which are regarded as "zinc thioneins" (Tio et al. 2004; Vasak and Meloni 2011). The molecular studies underlie the significance of MT-4 expression in the skin, as copper and zinc ions are important in modulating the development of normal skin, wound healing, and the progression of skin diseases (Kumar et al. 2012; Ala et al. 2013).

It has already been shown that MT-4 is involved in the regulation of the development of the stratified squamous epithelium of the skin and upper respiratory tract, which might support this thesis (Quaife et al. 1994). In situ hybridization has shown that the most intense *MT-4* mRNA signals are present in the differentiating spinous layer of cornified epithelia. In turn, the MT-1 isoform expression has been noted predominantly in the proliferative basal layer of epidermis (Liang et al. 1996). Downregulation of *MT-4* gene expression in nude mice characterized by the lack of Whn transcription factor activity has also been noted (Schlake and Boehm 2001). These observations point to a different role of MT-4 than of other members of the MT family and indicate its important role in the differentiation process of the skin (Quaife et al. 1994; Schlake and Boehm 2001). In recent studies the presence of MT-4 was not found in squamous cell cancers of the lung. Thus, it seems that the MT-4 role in differentiation may be limited only to the skin (Werynska et al. 2011).

In addition, MT-4 has also been identified in mouse maternal decidua (Liang et al. 1996). *MT-1* and *MT-2* mRNA expression have been found in the visceral yolk sac, placenta, and fetal liver; however, no *MT-3* or *MT-4* mRNA expression has been noted in these tissues. In turn, *MT-3* and *MT-4* mRNAs are both abundant in the maternal decidua and in experimentally induced deciduoma at 7 and 8 days *post coitum*, in which high levels of *MT-1* and *MT-2* mRNA have also been observed (Liang et al. 1996).

Chapter 3
Metallothionein-3

3.1 Introduction

Numerous studies on the MT-1 and MT-2 proteins are currently under way, but the MT-3 isoform, first described about 20 years ago by Uchida et al. as a neuronal growth-inhibitory factor (GIF), still raises many questions (Uchida et al. 1991). The original name was derived from the initially discovered functions of this protein, i.e., its strong ability to impair neurite overgrowth and neural survival of cultured neurons. Further studies on the structure and function of GIF showed a 70 % amino acid similarity to proteins of the MT family, which allowed it to be given its final name of MT-3 (Palmiter et al. 1992). MT-3 consists of 68 amino acids containing 20 cysteine residues in conserved positions with two inserts: an acidic hexapeptide in the C-terminal region and a threonin at position 5. Moreover, the conserved Cys-Pro-Cys-Pro motif between positions 6 and 9 is unique to MT-3 (Uchida et al. 1991; Romero-Isart and Vasak 2002; West et al. 2008). Primarily in physiological conditions, abundant MT-3 expression has been observed in fibrous and protoplasmic cortical astrocytes (Aschner 1996a, b). Subsequently, its expression on the mRNA and protein levels was disclosed in peripheral organs outside the brain, suggesting functions other than neurite-growth inhibition (Hozumi et al. 2008). Nevertheless, studies on the functions of the MT-3 in the nervous system dominate in the literature. Its mRNA and protein upregulation have often been observed following brain injury caused by various factors, whereas its downregulation has been observed in neurodegenerative conditions, such as Alzheimer's disease (AD). These findings directly suggest the strong involvement of MT-3 in central nervous system (CNS) repair (Chung et al. 2002; Lee et al. 2011; Manso et al. 2012; Luo et al. 2013). MT-3 shares several biological and chemical properties with other MT isoforms, mainly associated with their structure, e.g., the sequestration, distribution, and detoxification of metal ions. However, there are also very distinct chemical features that might be relevant to the particular biological functions of MT-3, such as its neuromodulation of catecholaminergic,

© Springer International Publishing Switzerland 2016
P. Dziegiel et al., *Metallothioneins in Normal and Cancer Cells*, Advances in Anatomy, Embryology and Cell Biology 218, DOI 10.1007/978-3-319-27472-0_3

glutamatergic, and GABAergic transmission in the nervous system (Frederickson and Moncrieff 1994).

It was initially suggested that MTs in the brain are involved in the regulation of metal homeostasis and neural protective function against toxic metals (Masters et al. 1994; Palmiter 1995). It has been found that MT-3 binds Zn and Cd ions less strongly than does MT-2, although the MT-3 gene contains metal-responsive elements in its promoter, which seems to be not inducible by heavy metals (Erickson et al. 1997). Due to the localization of the MT-3 protein, its expression has been studied in various pathological conditions, especially of the brain. Initially, MT-3 was found to exert growth-inhibitory activity on neurons cultured with AD brain extract (Uchida et al. 1991). The results suggested that the loss of MT-3 might result in elevated neurotrophic activity, inducing abortive sprouting exhaustion, causing neuronal death. This observation served as the point of reference for many experiments carried out in different neurodegenerative diseases, various forms of brain injuries, as well as pathological and physiological processes of aging brain (Sogawa et al. 2001; Thirumoorthy et al. 2011; Sharma and Ebadi 2014). These findings indicate that reductions in MT-3 may correlate with loss of neurons or of neuroprotection and that the exact roles of MT-3 are still unknown. Based on the known functions of the MT-1 and MT-2 proteins in neoplastic diseases, some studies concerning MT-3 expression in malignancies have also been conducted; however, results and conclusions are still very limited.

3.2 Structure of MT-3

MT-3 is a low molecular weight, heat-stable, cysteine-rich, acidic protein that possesses a high affinity for metal ions. The gene of human MT-3, like that of other members of the MT family, is found on chromosome 16 (16q12-22) (Karin et al. 1984). Human, mouse, and rat *MT-3* genes have been isolated and sequenced, and protein analysis has revealed that MT-3 consists of 68, 68, and 66 amino acid residues in these species, respectively (Uchida et al. 1991; Kobayashi et al. 1993; Naruse et al. 1994). MT-3 exhibits about 70 % sequence similarity with other well-studied MT isoforms (Palmiter et al. 1992). A characteristic feature of its structure is the length of the amino acid sequence, with 20 completely conserved cysteine residues and two domains, α and β (Uchida and Ihara 1995; Sewell et al. 1995). The total binding capacity of seven metal particles is provided by metal-thiolate clusters—three in the N-terminal β-domain $M(II)_3S_9$ and four in the C-terminal α-domain $M(II)_4S_{11}$. Moreover, two inserts have been distinguished: a threonine in position 5 and a glutamate-rich hexapeptide near the C-terminus, as a well-conserved CPCP (6–9) motif, which are distinctive of and unique to MT-3 (Cai et al. 2006; Ni et al. 2007). The acidic character of the molecule is provided by just these glutamic acid residues (Uchida et al. 1991). In the amino acid chain of MT-3, two proline particles (at positions 7 and 9) would seem to be essential for its role, but the bioactivity state is dependent not only on the sequence but also on the conformation (Sewell et al. 1995; Cai et al. 2006). It has been reported that the neuronal-inhibitory growth activity of MT-3 is mainly associated with its β-domain, whereas the α-domain is responsible for the stabilization of the proper structure (Romero-Isart and Vasak 2002; Zheng et al. 2003). It is worth mentioning that the linker region, localized between the α and β domains, is crucial for the modulation of the stability and accessibility of the β domain in particular and is thus indispensable to the biological activity of MT-3 (Ding et al. 2009). A determination of its proper structure is yet to be made, as crystallographic data on GIF are unavailable—probably because of a highly dynamic protein structure that is difficult to crystallize. Further studies have become possible through NMR techniques and molecular dynamic simulations (Oz et al. 2001; Wang et al. 2006).

3.3 Localization of MT-3

While the expression of MT-1 and MT-2 is activated by various stimuli, including drugs, metals, and inflammatory agents, and is widely expressed in almost all mammalian tissues, MT-3 is predominantly found within the CNS and is regulated differently than other MTs, without the metal-induced synthesis (Vasak and Meloni 2011). The specific localization of *MT-3* mRNA and protein in the CNS has been the subject of numerous studies, but confusing results have been obtained. There is evidence for the MT-3 transcript and protein being found only in astrocytes (Kobayashi et al. 1993; Nakajima and Suzuki 1995; Hozumi et al. 1998; Ballestín et al. 2014), only in neurons (Masters et al. 1994; Choudhuri et al. 1995; Nordberg 1998; Gong and Elliott 2000), as well as in both glial and nerve cells (Anezaki et al. 1995; Kramer et al. 1996; Yamada et al. 1996). In analyzing the function of MT-3, despite the numerous contradictory data, one common assumption is that the MT-3 protein is primarily synthesized in the glial cells (mostly in astrocytes), being predominantly found in the cell body and its fine processes. Astrocytic expression of MT-3 has been found in all the major regions of the brain—in the cortex, brainstem, and spinal cord (Choudhuri et al. 1995). It should be noted that neuronal expression of MT-3 has also been observed at a much lower level in these CNS areas. The neuronal expression of MT-3 has been found to dominate in other locations—mostly in those with high Zn concentrations in synaptic vesicles, such as the hippocampus (predominantly the dentate gyrus), the piriform cortex, the amygdala, the olfactory bulb, and the Purkinje cells of the cerebellum (Masters et al. 1994).

There are many lines of evidence documenting the expression of MT-3 in several peripheral organs. The study conducted by Moffatt et al. showed that MT-3 mRNA is expressed in the testes, prostate, epididymis, tongue, ovary, uterus, stomach, heart, and seminal vesicles of the rat (Moffatt and Séguin 1998). MT-3 expression on the mRNA and protein levels has been well documented in the glomeruli and the collective tubules of the kidney, prostate glandular epithelium, Sertoli cells, and Leydig cells of the testes, and in the taste buds in the tongue

(Moffatt and Séguin 1998). In humans, MT-3 expression was initially detected in the kidney, with a particular localization in the cytoplasm of the distal tubular cells (Hoey et al. 1997; Garrett et al. 1999). The physiological expression of MT-3 has been demonstrated in some cells of the male genitourinary tract—heterogeneously in the Leydig cells and seminiferous tubules of the testes on the mRNA level— whereas the MT-3 protein has been found in the prostate epithelial and stromal cells (Moffatt and Séguin 1998; Garrett et al. 1999). Despite the rare occurrence of MT-3 outside the CNS, it should be noted that its presence may carry additional functions, mostly in view of neoplastic diseases of these organs.

3.4   Functions of MT-3 in the Brain

Dysregulation of metal-ion homeostasis seems to be one of the most important aspects of the etiopathogenesis of neurodegenerative diseases (Bonda et al. 2011). In nervous tissue, many different mechanisms are physiologically activated to prevent neurodegeneration, mostly in regard to the course of normal aging, but also triggered by spontaneous actions such as injury or intoxication (Bonda et al. 2011). Generally, all MT isoforms, with the exception of MT-4, can be found in nervous tissue. MT-1 and MT-2 induce cell growth and differentiation, whereas MT-3 inhibits proliferation of nerve and glial cells. Typically, in neuro-degenerative processes, decreased levels of MTs are observed, resulting in early aging and increased morbidity and mortality. It has been suggested that MTs have neuroprotective potential, especially in their ability to sequester and disperse metal ions, to control metalloproteinases' activity (e.g., Zn-dependent transcription factors), and to protect the cells from the deleterious action of reactive oxygen species, ionizing radiation, anticancer drugs, and mutagens (Aschner et al. 2006).

Initially, MT-3 was considered a GIF, and in nervous tissue it was found to play an important role such as inhibiting the survival and neurite formation of neurons in cell cultures (Uchida et al. 1991). Subsequent studies conducted by Hozumi et al. showed upregulation of MT-3 mRNA and protein in induced brain injuries (Hozumi et al. 1998). In view of the downregulated MT-3 observations in Alzheimer's disease (AD) patients, MT-3 involvement in the repair of CNS has been suggested (Yu et al. 2001; Colangelo et al. 2002). Many studies have shown that neither MT-3 mRNA nor protein levels are significantly decreased in AD (Uchida et al. 1991; Yu et al. 2001; Colangelo et al. 2002), whilst Carrasco et al. have reported increased level of MT-3 protein in the brains of patients with AD, compared to healthy patients (Carrasco et al. 1999). Moreover, an increased level of MT-3 was found in reactive astrocytes in the cerebral cortex during meningitis and Creutzfeldt–Jakob Disease (CJD), as well as in reactive astrocytes surrounding old cerebral infarcts (Uchida 1994). On the other hand, decreased expression of MT-3 was demonstrated in the CJD cortex, and the authors suggested that the MT level could vary depending on the duration of the disease (Kawashima et al. 2000). The reduced level of MT-3 was also seen in neurological disorders, such as multiple system atrophy, Parkinson's disease (PD), progressive supranuclear palsy, and amyotrophic lateral sclerosis (ALS) (Ono et al. 2009; Hozumi 2013; D'Amico et al. 2013). In brains of elderly Down Syndrome (DS) patients with Alzheimer-type dementia, MT-3 was decreased in astrocytes around senile plaques, which may be correlated with neuronal loss or degeneration and indirectly explain why DS predisposes to premature dementia (Arai et al. 1997). However, in vitro studies provided opposite results. The murine model of DS cell line Ts65Dn displayed reduced intracellular Zn levels and increased expression of MT-3 (Ballestín et al. 2014).

The colocalization of MT-3 and Zn in neurons suggests several possible unique functions of MT-3. Masters et al. have suggested that MT-3 participates in the utilization of Zn as a neuromodulator (Masters et al. 1994; Lee et al. 2010). Like other members of the MT family, MT-3 might play a role in the detoxification of metal ions and facilitating their transport, and protecting cells from damage, as also noticed in a number of neurodegenerative diseases (Lee et al. 2010). It has been clearly shown that MT-3 is not essential in protecting against exogenous heavy metal ions, because MT-3 knockout mice were not more sensitive than normal mice to Zn or Cd toxicity (Masters et al. 1994). On the other hand, it has been shown that MT-3 knockout mice are highly sensitive to kainate-induced seizures, which are associated with glutaminergic neurotransmission (Erickson et al. 1997). During the seizures, Zn and glutamate are released from synaptic vesicles and activate kainic and glutaminergic receptors on postsynaptic neurons in a synergistic manner. After release, Zn is probably recovered by specific uptake mechanisms and recycled into the synaptic vesicles. Aschner et al. have suggested that MT-3 might play a role in this recycling process (Aschner et al. 1996b).

Initially, only MT-1 and MT-2 were suspected to scavenge free radicals and protect cells against oxidative stress (Thornalley and Vasak 1985; Sato and Bremner 1993). However, further in vitro studies on MT-3 have also shown its high efficiency in cell protection (Uchida 1994; Uchida and Ihara 1995). Many studies have proposed that the increased expression of MT-3 induced by reactive oxygen species may protect cells against oxidative stress (Sogawa et al. 2001; Lee et al. 2014). A possible pathway is provided by Cab1 and Mst1 kinases regulated by changes in levels of free zinc ions released from MT-3 and resulting in oxidative cell death (Lee et al. 2014). However, at a later stage of neurodegenerative disease, or even of brain injury, neuronal damage may disrupt normal neuroglial interactions and, consequently, MT-3 synthesis may be reduced in reactive astrocytes (Uchida 1994; Hozumi et al. 1998).

It seems to be the case that, in line with the primary observed function of MT-3, its synthesis by reactive astrocytes and subsequent secretion (under certain conditions) could be a mechanism of the brain's response to stressful conditions, such as physical or chemical trauma. Howells et al. proposed a possible model of MT-3 involvement in the cellular response to brain injury. They postulated that, during neuronal degeneration, adjacent glial cells respond by presenting reactive phenotype and proliferation. MT-3 produced by astrocytes may primarily decrease neurite overgrowth as a response to the injury. Subsequently, the synthesis of

MT-3 declines as a result of disturbed communication between neurons and glia. Moreover, in advanced stages of neurodegenerative diseases, the synthesis and release of MT-3 may be decreased as a result of neuronal damage. Finally, reduced MT-3 levels may be responsible for the suppression of cellular antioxidant defenses, followed by the impairment of cell proliferation (Howells et al. 2010). Such a mechanism, being of universal character, might be extrapolated to a wider spectrum of injuries, mostly caused by ischemic stress and reactive oxygen species (Yuguchi et al. 1995, 1997).

3.5   MT-3 Expression in Cancer

The presence of MT-3 outside the CNS was first revealed in parts of the urogenital tract. The MT-3 protein has been found in particular in the kidney, prostate, and urinary bladder, in both normal and tumor-affected cells (Hoey et al. 1997; Garrett et al. 1999). Only one work has shown a correlation between the expression of MT-3 and an increase in the grade of histological malignancy (G). The authors of that study concluded that the upregulation of MT-3 expression can potentially be used as a marker of malignancy in bladder cancer (Sens et al. 2000). However, it is difficult to explain the increased expression of MT-3 in tumor cells—a protein presenting antiproliferative activity. It has been hypothesized that an increase in apo-MT induced by zinc deficiency could result in the dysregulation of the cellular metabolism and genome instability (Cherian et al. 2003). Another study investigated MT-3 expression in breast cancer, finding that its overexpression was associated with poor prognosis, concomitantly with others, who postulated that MT-3 absence was a favorable marker (Sens et al. 2001). Moreover, it was also suggested that this phenomenon was under epigenetic control (Somji et al. 2010). Increased expression of MT-3 in neoplastic lesions, compared to normal tissues, was found in lung and urinary bladder cancers (Somji et al. 2011; Werynska et al. 2013a, b). In the PC-3 prostate cancer cell line, the overexpression of *MT-3* results in growth inhibition and induction of chemotherapeutic drug resistance (Dutta et al. 2002). On the other hand, it was found that *MT-3* expression was downregulated by androgen in the LNCaP prostate cell line (Otsuka et al. 2013). Androgen deprivation (standard in prostate cancer treatment) induces upregulation of MT expression and subsequently a decrease in Zn deposits, which is a typical risk factor for neoplasms originating from the urothelium (Otsuka et al. 2013). In the gastrointestinal tract, downregulation of MT-3 has been observed mainly in the esophagus and stomach (Smith et al. 2005). The decreased expression of MT-3 was found in esophageal squamous cell carcinomas and was shown to result from DNA methylation (Smith et al. 2005). Hypermethylation of the CpG island of the *MT-3* gene region was associated with this downregulation and resulted in the silencing of an *MT-3* gene in gastric and esophageal cancer (Deng et al. 2003; Peng et al. 2011). Two esophageal cancer cell lines (Eca-109 and TEI3) transfected by *MT-3* show inhibition of proliferation and increased rate of cells undergoing apoptosis (Tian et al. 2013).

The authors suppose that transfection technique might have a therapeutic potential in this type of cancer. However, this effect was found to be limited. A well-known fact is that MT expression increases in tumor cells resistant to chemotherapy, including cisplatin and its derivatives. It was shown that MT-3 has a much faster reaction to platin-derivative drugs than MT-2 and might be responsible for the acquired resistance to chemotherapy in MT-3 positive cells (Karotki and Vasak 2009). In a recent study, Pula et al. analyzed the expression of MT-3 in normal skin and its malignancies. A significantly higher level of MT-3 was observed in squamous cell cancer, which suggests the association between MT-3 expression and development of this type of skin cancer (Pula et al. 2015).

In view of the role of the MT-3 protein, it can be stated that in neoplastic diseases mostly decreases in its expression were observed. However, in designing therapeutic strategies, it is assumed that MT-3 has the potential to induce apoptosis as well as inhibit cell growth. Thus, further studies on the role of MT-3 in cancerogenesis and cancer prediction are warranted.

Chapter 4
The Role of Metallothioneins in Carcinogenesis

4.1 Introduction

Due to their role in various cellular processes, metallothioneins have been studied in numerous benign and malignant lesions, in regard to carcinogenesis, and in tumor progression (Dziegiel 2004; Krizkova et al. 2009b, 2012; Pedersen et al. 2009; Pula et al. 2012; Fic et al. 2013). Although the role of MTs in tumor initiation and progression has been intensely studied, it is only recently that studies have allowed the elucidation of the tumor suppressory roles of some MT isoforms—such as of MT-1G in papillary thyroid cancer and the large intestine and MT-1E in melanomas (Ferrario et al. 2008; Faller et al. 2010; Arriaga et al. 2014). On the other hand, MT-2A has been shown to promote the progression of breast and non-small cell lung cancer (Jin et al. 2002; Lim et al. 2009; Lai et al. 2010; Werynska et al. 2013b). Lines of evidence indicate that MTs exert divergent expression patterns in normal tissue and tumors; for this reason, more detailed studies concentrated on each isoform, and not on MT-1 and MT-2 collectively, need to be conducted. Nevertheless, the results of many studies have identified the involvement of MTs in cancer cell proliferation (Jin et al. 2002), apoptosis (Krizkova et al. 2009b), invasiveness, and migration (Jin et al. 2001; Kim et al. 2011; Ryu et al. 2012), as well as on cancer cells sensitivity to various anticancer agents (Kondo et al. 1995a, b, 1997). In this chapter, the role of MT isoforms in the development and progression of various tumors, with special emphasis on their prognostic significance, will be summarized in an organ-specific manner.

4.2 Hematological Malignancies

The expression of MT in hematopoietic cells has been widely described, yet many contrary results have been published. It has been shown that MTs play a role in hematopoietic stem cell differentiation (Maghdooni Bagheri and De Ley 2011; Takahashi 2012), in proliferation (Abdel-Mageed et al. 2003; Duval et al. 2006), and in the prevention of apoptosis (Tsangaris and Tzortzatou-Stathopoulou 1998; Takahashi 2012). Although many authors analyzed potential usefulness of MT expression as a hematological malignancy marker, so far the only evident role appears to be their relation to drug resistance in acute myeloid leukemia (AML) (Sauerbrey et al. 1998) and acute lymphoid leukemia (ALL) (Tsangaris et al. 2000). MTs were shown to be a poor prognostic marker in diffuse large B-cell lymphomas (DLBCL) (Poulsen et al. 2006). As there are many subtypes of hematological malignancy that occur predominantly in childhood, investigations were also conducted on pediatric patients.

Sauerbrey et al. studied the expression of MT-1/2 in initial and relapsed childhood ALL. No significant differences between either initial or relapsed ALL and MT-1/2 expression were shown in regard to disease-free survival (Sauerbrey et al. 1998). Conversely, other studies demonstrated that the expression of MT-1/2 correlated with chemoresistance, possibly being a sensitive antiapoptotic factor (Tsangaris et al. 2000). The study conducted by Usvasalo et al. using microarray comparative genomic hybridization (CGH) demonstrated that gene sets in young ALL patients differed between those who had a risk of relapse and those who did not. The prognostic classifier set consisted of BCL2-antagonist/killer 1 (*BAK1*), cyclin-dependent kinase inhibitor 2 C (*CDKN2C*), glutathione-*S*-transferase M1 (*GSTM1*), and *MT-1F* (Usvasalo et al. 2010). From a clinical point of view, knowledge of chemoresistance is crucial since chemotherapy is the treatment of choice in hematological malignancies. *The expression of* resistance-related proteins P-glycoprotein 170 (P-170), glutathione-*S*-transferase pi (GST-Pi), topoisomerase-II (Topo II), thymidylate synthase (TS), and MT-1/2 was investigated in 19 AML children (Sauerbrey et al. 1998). It was found that patients who developed relapse showed a poorer prognosis and frequently expressed more than two resistance-related proteins, including MT-1/2, as compared with patients who remained in remission (Sauerbrey et al. 1998). It has been recently reported that the mRNA expressions of the *MT-1A* and *MT-1G* genes were negatively correlated with the *PU.1* gene (a gene controlling hematopoiesis) in 43 primary AML specimens (Imoto et al. 2010). The authors did not analyze the clinical outcomes of these patients, but it was previously shown that *PU.1* expression correlated inversely with the tyrosine kinase receptor FLT3 (Inomata et al. 2006), whose strong expression was shown to be an unfavorable prognostic factor for overall survival (Kuchenbauer et al. 2005). Other reports also suggest that *PU.1* expression is a positive indicator for other hematological malignancies, such as follicular lymphoma (Torlakovic et al. 2006).

There are some suggestions that the expression profiles of MT-1/2 may be discriminative for malignant and nonmalignant neoplasms of lymphoid origin (Rizkalla and Cherian 1997; Poulsen et al. 2006). Hodgkin lymphoma (HL) consists of two distinct entities: nodular-lymphocyte predominant HL (NLPHL) and classical HL (CHL). CHL includes four histological subtypes: nodular sclerosis HL (NSHL), mixed cellularity HL (MCHL), lymphocyte-rich classical HL (LRCHL), and lymphocyte-depleted classical HL (LDHL) (Tzankov and Dirnhofer 2006). The MT expression was analyzed in the lymph nodes of 34 patients with HL (NSHL, MCHL, LRCHL, and NLPHL patients), looking for some group-discriminating results. Strong MT-1/2 expression was observed in the lymph nodes of NSHL and MCHL patients, in contrast to LRCHL patients who barely expressed MT-1/2. In NLPHL patients, a heterogeneous pattern of MT-1/2 expression was observed. The authors stated that MT-1/2 was differentially expressed in HL subclasses (Penkowa et al. 2009), but no therapeutically useful information was gained, as there were no correlations between the clinical outcomes after chemotherapy and the expression of MT-1/2 of the patients with HL. Interestingly, Poulsen et al. reported a significant inverse correlation between MT-1/2 expression and clinical outcome in DLBCL (Poulsen et al. 2006). The MT mRNA was upregulated in 15 of 48 DLBCL patients, but only one of 15 patients with upregulated MT mRNA achieved sustained remission. This strongly suggests that upregulation of MT mRNA significantly increases the risk of treatment miscarriage (Poulsen et al. 2006). For confirmation of their results, Poulsen et al. conducted an immunohistochemical analysis, observing a significantly poorer 5-year survival, independent of age, stage, or international prognostic index in relation to increased expression of MT-1/2 (Poulsen et al. 2006). Wrobel et al. studied MT-1/2 and GST-Pi expression in the bone marrow of patients with myeloproliferative disease (MPD). Although the examined group consisted of only nine patients with osteomyelofibrosis (OMF) and 11 patients with chronic myelocytic leukemia (CML), increased expressions of both tested markers were found. The MT-1/2 expression was stronger in OMF patients and associated with bone marrow fibrosis (Wróbel et al. 2004).

4.3 The Skin and Nonmelanoma Skin Tumors

The role of MT-1/2 expression in the skin has been best described by McGee et al., who summarized the divergent effects of its expression in the skin (McGee et al. 2010). In normal skin, MT-1/2 immunoreactivity is present in the proliferating basal layer of the epidermis (Fig. 4.1a) (Zamirska et al. 2012), where it has been shown to protect against ultraviolet radiation B (UV-B), *cis*-urocanic acid, and cholera toxin in both experimental animals and humans (Hanada et al. 1998; Hanada 2000; Nishimura et al. 2000; Reeve et al. 2000; Ablett et al. 2003). The photoprotective effect of isoflavonoids seems to be MT dependent, as MT-null mice show no suppression of protection upon isoflavonoids administration (Widyarini

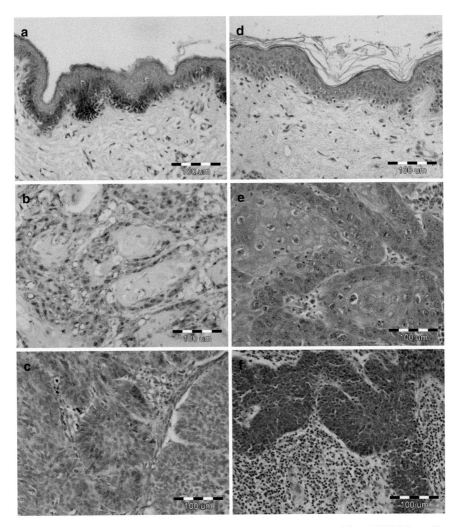

Fig. 4.1 *Immunohistochemical demonstration of metallothionein (MT) 1/2 and MT 3 in a skin cancer.* Cytoplasmic expression of MT 1/2 (**a**) and MT 3 (**d**) in normal keratinizing squamous stratified epithelium (MT 1/2 expression observed predominantly in basal layer). Strong cytoplasmic expression of MT 1/2 can be observed in squamous cell skin cancer (**b**) and in basal cell skin cancer (**c**) and relatively weaker expression of MT3 in squamous cell skin cancer (**e**) and in basal cell skin cancer (**f**). Archival sections from the Department of Histology and Embryology, Wroclaw Medical University, Wroclaw, Poland

et al. 2006). The immunoreactivity of MT-1 and MT-2 has also been shown to be upregulated in epidermis during the healing of experimentally induced wounds (Iwata et al. 1999). Furthermore, the endogenous expression of MT-1/2 also increased in basal keratinocytes concurrent with reepithelialization after a burn injury (Morellini et al. 2008). Human keratinocytes challenged in vitro with exogenous Zn(7)-MT-2A are characterized by increased cell viability and migration abilities, as well as decreased apoptosis (Morellini et al. 2008). In contrast, immortalized primary cultured skin fibroblasts were not responsive to Zn(7)-MT-2A as no significant changes in cellular migration and contraction could be observed. Thus, the expression of MT-1/2 in keratinocytes is important for accelerated wound healing (Morellini et al. 2008). The protective effects of MT-1/2 seem to decrease with advanced patients' age and chronic sun exposure. The MT-1/2 protein levels decreased significantly with aging in sun-protected and sun-exposed skin; however, the latter presented significantly lower MT-1/2 protein levels (Ma et al. 2011). The lower levels in the sun-exposed and aged skin could partially explain the frequent occurrence of nonmelanoma skin cancers in elderly people due to lower intensity of ROS scavenging.

Indeed, lines of evidence suggest that MT-1/2 could prevent skin carcinogenesis in animal models. MT-null mice have been shown to be more susceptible to 12-dimethylbenz[a]anthracene (DMBA) and to combined DMBA and 12-O-tetradecanoylphorbol-13-acetate (TPA) intoxication than wild-type mice (Zhang et al. 1998; Suzuki et al. 2003). However, once the carcinogenesis process occurs, MT-1/2 loses its protective functions and promotes tumor growth (McGee et al. 2010). In nonmelanocytic skin lesions MT-1/2 expression increases with the malignancy of lesions: the lowest levels are noted in actinic keratosis (carcinoma in situ), moderate levels in basal cell carcinomas (BCC), and the highest levels are observed in squamous cell carcinomas (SCC) (Fig. 4.1b, c) (Borges Junior et al. 2007; Zamirska et al. 2012). MT-1/2 immunoreactivity correlates positively with the expression of the Ki-67 antigen in actinic keratosis, SCC (Zamirska et al. 2012), and BCC (Bieniek et al. 2012), suggesting a role for MT-1/2 in regulating tumor cell proliferation rate. Nevertheless, the impact of MT-1/2 isoform expression on the survival of patients with such tumors remains to be determined.

Recently, low cytoplasmic MT-3 expression has been observed in normal skin epidermis with faint or no expression in the epidermal basal layer (Fig. 4.1d) (Pula et al. 2015). The increased expression of MT-3 has been observed in actinic keratosis and SCC cases, while its expression remained low in BCC cases (Fig. 4.1e, f). The results of this study indicate that MT-3 could be used as a marker of squamous differentiation in malignant tumor of the skin (Pula et al. 2015).

4.4 Melanoma

The expression of MT-1 and MT-2 has also been studied in cutaneous malignant melanomas, where their overexpression is associated with more rapid disease recurrence, metastatic spread, and poor patient outcome (Weinlich et al. 2003, 2006, 2007; Weinlich and Zelger 2007; Emri et al. 2013). Furthermore, overexpression of MT-1/2 in melanoma cells is correlated with the count of tumor-infiltrating CD68-expressing macrophages—high numbers of which are regarded as a poor prognostic factor for this malignancy. This indicates that

MT-1/2 may play a role in a defective tumor–host response (Emri et al. 2013). Molecular studies identified *MT-1E* as a potent oncosuppressor of malignant melanoma (Faller et al. 2010). Its expression decreases due to promoter hypermethylation in this malignancy, as compared with benign melanocytic nevi. Moreover, metastatic lesions present decreased levels of *MT-1E* mRNA more frequent than primary malignant melanomas (Faller et al. 2010). Furthermore, the ectopic overexpression of *MT-1E* in melanoma cell lines results in their diminished survival, as compared with control cells subjected to cisplatin treatment (Faller et al. 2010).

4.5 Mammary Gland and Breast Cancer

To date, the expression of particular MT isoforms has been best studied in breast tissues and its cancers. In the normal breast and its benign lesions (e.g., adenosis, scleroadenosis, and papillomas), MT-1/2 expression has been observed in myoepithelial cells surrounding the ducts and acini; their expression has rarely been noted in luminal epithelial cells (Fig. 4.2) (Bier et al. 1994). However, in epitheliosis, both the luminal and myoepithelial cells expressed MT-1/2. Ductal breast carcinomas in situ (DCIS) and invasive ductal breast carcinomas (IDC) are characterized by high MT-1/2 expression, compared to their lobular equivalents—the lobular carcinoma in situ (LCIS), the invasive lobular carcinoma (ILC), and the lobular hyperplasia (Fig. 4.3) (Bier et al. 1994; Douglas-Jones et al. 1995; Gallicchio et al. 2004). MT-1/2 immunoreactivity in cancer cells was higher in DCIS cases with comedo-like necrosis, compared with such cases without necrosis (Douglas-Jones et al. 1995). Moreover, MT-1/2 expression increased with the cytological grade of DCIS (Douglas-Jones et al. 1995). Alterations in MT-1 expression have also been noted in tumor adjacent tissue, where its elevated expression could be noted particularly in cases of breast cancer with confirmed lymph node metastases (Dutsch-Wicherek et al. 2005). MT-3 has also been shown to be overexpressed in breast cancers; however, its expression has been not noted in normal ductal epithelium (Sens et al. 2001; Somji et al. 2010). This observation has been confirmed by in vitro experiments in which the MCF-10A cell line—an immortalized nontumorigenic model of human breast epithelial cells—presented with no basal expression of MT-3, which could also not be induced by cadmium treatment. Demethylation agent (5-aza-2′-deoxycytidine) and histone deacetylase inhibitor (MS-275) increased *MT-3* mRNA expression probably at the level of MT-3 promoter (Somji et al. 2010). MT-3 was diffusely localized to the cytoplasm of IDC cancer cells, and its high expression showed a trend toward poor patient outcomes (Sens et al. 2001; Somji et al. 2010). Interestingly, when the analysis was limited to DCIS, the intensity at MT-3 staining was significantly increased in patients with poor prognosis in relation to those with good prognosis (Sens et al. 2001). Moreover, increased nuclear MT-3 expression in triple-negative breast cancer cases was found to be associated with patients' shorter disease-specific

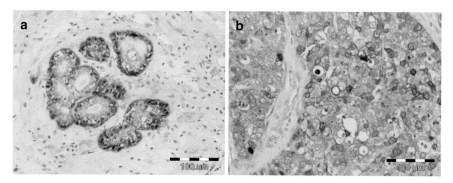

Fig. 4.2 *Immunohistochemical demonstration of metallothionein 1/2 in cytoplasm and nuclei of myoepithelial cells of normal breast tissue* (**a**) *and invasive ductal breast cancer cells* (**b**). Archival sections from the Department of Histology and Embryology, Wroclaw Medical University, Wroclaw, Poland

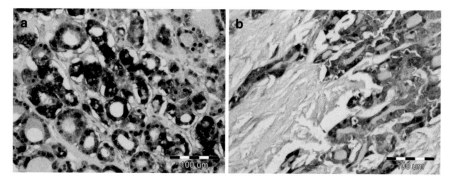

Fig. 4.3 *Immunohistochemical demonstration of metallothionein 1/2 in cytoplasm and nuclei of follicular* (**a**) *and medullary* (**b**) *thyroid carcinoma.* Archival sections from the Department of Histology and Embryology, Wroclaw Medical University, Wroclaw, Poland

survival (Kmiecik et al. 2015). However, other in vitro experiments seem to contradict the clinical observations, as MT-3 inhibited the growth of MCF-7 and Hs578T breast cancer cells (Gurel et al. 2003).

MT-1/2 immunoreactivity in breast cancers, in particular in IDC, has been intensely studied in regard to patients' clinical and pathological data; significant correlations have been reported in some of the studies. In the majority of the cases, MT-1/2 expression correlated positively with the malignancy grade of the tumors (Haerslev et al. 1994; Zhang et al. 2000; El Sharkawy and Farrag 2008; Gomulkiewicz et al. 2010; Wojnar et al. 2010, 2011). High MT-1/2 cancer cell expression levels have been shown to be associated with larger primary tumor size, the presence of lymph node metastases, a high number of mitoses, and the absence of progesterone receptors in a single study (Haerslev et al. 1995). However, many other studies have failed to confirm these clinical and pathological findings (Goulding et al. 1995). In the majority of cases, the expression status of estrogen and progesterone receptors of the analyzed breast cancers was inversely correlated with MT-1/2 immunoreactivity (Haerslev et al. 1995; Oyama et al. 1996; Ioachim et al. 1999, 2003; El Sharkawy and Farrag 2008; Gomulkiewicz et al. 2010; Wojnar et al. 2011). Furthermore, in some of the studies, high MT-1/2 expression in cancer cells was associated with poor outcome for IDC patients, as shown in univariate and multivariate survival analyses (Goulding et al. 1995; Haerslev et al. 1995; Vazquez-Ramirez et al. 2000; Zhang et al. 2000; Surowiak et al. 2006). Elevated MT-1/2 immunoreactivity in IDC also predicts tamoxifen resistance (Surowiak et al. 2005). Prognostic significance has also been studied in regard to MT-1/2 expression in IDC myoepithelial cells; however, no impact on patient survival could be observed (Surowiak et al. 2002).

Since these studies were performed using antibodies that could not distinguish particular MT-1/2 isoforms, additional molecular and in vitro experiments were conducted to elucidate the functions of these molecules in breast cancer. Using RT-PCR, *MT-1A*, *MT-1E*, *MT-1F*, *MT-1G*, *MT-1H*, *MT-1X*, and *MT-2A* (but not *MT-1B*) mRNAs have been detected in breast cancer tissue samples (Jin et al. 2002). So far, the *MT-2A* isoform has been identified as the most abundant in breast cancer. Its high mRNA expression correlates with MT-1/2 immunoreactivity and Ki-67 antigen expression, indicating its possible regulatory role in cellular proliferation (Jin et al. 2002). Positive correlations of MT-1/2 expression with proliferation markers (Ki-67 antigen and minichromosome maintenance protein-2, MCM2) have been noted in IDC cases (Gomulkiewicz et al. 2010; Wojnar et al. 2010, 2011). Furthermore, the silencing of *MT-2A* expression in MCF-7 decreased growth and induced cell-cycle arrest in the G1 phase (G1 arrest) and a marginal increase in cells in the sub-G1 phase (Lim et al. 2009). Concomitantly, higher expression of the *ataxia telangiectasia mutated (ATM)* gene and the lower expression of the *cdc25A* gene following *MT-2A* silencing has been observed—which points to the involvement of this isoform in the regulation of the ATM/Chk2/cdc25A pathway (Lim et al. 2009). Furthermore, the *MT-2A* isoform has been shown to regulate entosis of MCF-7 cells; thus, in addition to its cell-cycle regulatory capabilities, a high expression of *MT-2A* could protect cancer cells from this type of death (Lai et al. 2010). The overexpression of *MT-2A* in invasive MDA-MB-231 breast cancer cells induced invasiveness, but its silencing completely inhibited both cell invasion and migration (Kim et al. 2011). *MT-2A* gene expression levels have been found to be higher in invasive MDA-MB-231 than in noninvasive MCF-7 cells (Tai et al. 2003). Moreover, the overexpression of *MT-2A* promoted MDA-MB-231 cell invasion by upregulating MMP-9 via activation of AP-1 and *NF-kB* (Kim et al. 2011). The silencing of *MT-2A* also led to increased chemosensitivity to doxorubicin in MCF-7 breast cancer cells (Yap et al. 2009). Based on these findings, the impact of *MT-2A* gene polymorphism on increased breast cancer risk was extensively studied (Seibold et al. 2011, 2013; Krzeslak et al. 2014).

A number of molecular studies have permitted identification of the potential roles of other MT-1/2 isoforms in breast cancer. *MT-1F* mRNA has been positively correlated with the malignancy grade of IDC cases (Jin et al. 2001), whereas the expression of *MT-1E* mRNA has been found in estrogen receptor negative IDC specimens and cell lines (Friedline et al. 1998; Jin et al. 2000). Furthermore, increased expressions of *MT-1E*, *MT-1X*, and *MT-2A* have been shown to be associated with the metal toxicity resistance of PMC42, MCF-7, and MDA-MB-231 breast cancer cells (Barnes et al. 2000; Alonso-Gonzalez et al. 2008). Although other MT-1/2 and MT-3 isoforms seem to possess protumorigenic potential, the *MT-1G* isoform seems to function as a potential oncosuppressor, as its expression has been found to be downregulated in breast cancer, compared with normal breast tissues (Friedline et al. 1998; Park et al. 2011).

4.6   Thyroid Gland

Metallothionein expression has been intensely studied in both benign and malignant thyroid lesions. The first report of Nartey et al. revealed the cytoplasmic–nuclear immunoreactivity of MT-1/2 utilizing rabbit polyclonal antibody in normal and tumor cells of the thyroid (Nartey et al. 1987a). In this pilot study, expression of MT-1/2 was detected in 91 % of the surgically resected thyroid tumors, as compared to 20 % of the normal thyroids sampled during autopsy procedures (Nartey et al. 1987a). However, later studies did not confirm these observations, as differential MT-1/2 expression has been noted in thyroid tumors (Huang et al. 2001, 2003; Finn et al. 2007; Ferrario et al. 2008; Krolicka et al. 2010) (Fig. 4.3). By use of molecular methods, such as expression microarrays and real-time PCR, it has been revealed that the expression of MT isoforms is downregulated in papillary thyroid carcinomas (PTC) and follicular thyroid carcinomas (FTC), as compared to normal thyroid (Huang et al. 2001, 2003; Finn et al. 2007; Pula et al. 2012).

The MT-1G isoform has been identified as a potent oncosuppressor of PTC (Huang et al. 2003; Ferrario et al. 2008; Fu et al. 2013). Its decreased mRNA expression has been noted in PTC tissues, as well as in cell lines derived from human well-differentiated (K1 and K2) and poorly differentiated (NPA-87) PTC (Huang et al. 2003). Besides *MT-1G*, *MT-1H* and *MT-1X* have also been identified as being downregulated in PTC tissues (Ferrario et al. 2008). It has been shown that hypermethylation of the promoter regions of *MT-1G*, but not the loss of heterozygosity (LOH), is responsible for the downregulated expression. Moreover, the treatment of K2 cells with 5-aza-2′ deoxycytidine (5-aza-dC; a demethylating agent) and trichostatin A (a histone deacetylase inhibitor) resulted in increased *MT-1G* expression (Huang et al. 2003). The restoration of MT-1G expression in K1 cells with the use of *MT-1G*-myc expression plasmid decreased proliferation rate and diminished their colony formation potential as compared to the control cells.

The suppressive role of this MT was supported also by experiments in which the *MT-1G*-overexpressing K1 cells, upon injection to athymic mice, were characterized by reduced tumor growth. This was also confirmed in vitro (Ferrario et al. 2008). Moreover, *MT-1G* promoter hypermethylation has been found in about 30 % of thyroid cancers overall, with an increase noted in PTC (Fu et al. 2013). Other thyroid malignancies, such as FTC, MTC (medullary thyroid carcinoma), and ATC (anaplastic thyroid carcinoma), were characterized by respective lower *MT-1G* methylation levels. The hypermethylation of its promoters has also been detected in nodular goiter (NG) (Fu et al. 2013). Hypermethylation and histone modifications have also been found to be responsible for the downregulation of *MT-1G* expression in thyroid cancer cell lines (BCPAP, FTC133, IHH4, K1, 8305C, C643), whereas cell lines derived from normal thyroid epithelial cells showed minimal promoter methylation (Fu et al. 2013). Cell transfected with an *MT-1G* overexpression vector manifested decreased proliferation rates, particularly the K1 and FT133 cells, as compared to the cells transfected with control blank vectors. Moreover, these cells were characterized by a cell-cycle arrest, increased apoptosis rate, and decreased migratory capabilities (Fu et al. 2013). They also formed fewer and smaller colonies than the control cells. The K1 cells also presented decreased invasiveness following *MT-1G* overexpression. In the FTC133 cells, the migration of cells in the wound scratch assay was markedly inhibited in the *MT-1G*-transfected cells as compared to empty vector-transfected cells (Fu et al. 2013). Presumably, MT-1G may act in the thyroid as a tumor suppressor by modulating the activity of the PI3K/Akt pathway (Fu et al. 2013). It has been shown that experimental MT-1G overexpression induces the stability of the p53 protein and its downstream targets (p21, Bak, and Smac) and decreases the expression of Mdm2 in the K1 cells, though not in the FTC133 cells, probably due to the presence of the p53 mutation (Fu et al. 2013). Moreover, *MT-1G* overexpression has been shown to increase the expression of E-cadherin, but not to modify the expression of markers regarded as indicators of the epithelial-to-mesenchymal transition (EMT)—e.g., *Vimentin, Snail, Slug,* and *Twist.* MT-1G also decreased the phosphorylation of the Rb protein, indicating that the observed diminution in the proliferation rate of cancer cell lines may also be partially mediated by the inhibition of the Rb/E2F pathway (Fu et al. 2013). These in vitro results have also been confirmed by clinical observations, as PTC cases characterized by *MT-1G* promoter hypermethylation were more likely to develop lymph node metastases (Fu et al. 2013).

MT expression in tumor cells may be regulated by several factors. It has been shown that stimulation with thyroid-stimulating hormone receptor (TSHR) induces the expression of *MT-1X* in human FTC cells (Back et al. 2013). In these cells, the induction of *MT-1X* expression critically relies on intact Gq/11 signaling of the TSHR and is blocked by chelation of intracellular calcium and inhibition of protein kinase C (PKC) (Back et al. 2013). However, the inhibition of protein kinase A (PKA) does not affect the regulation of MT-1X by TSH, which could indicate the minimal role of this signaling pathway (Back et al. 2013). Cadmium ions have also been shown to increase MT-1/2 expression in PTC cancer cells (KAT5 cell line) and ATC cells (ARO cell line) (Liu et al. 2007, 2009). Following treatment with cadmium ions, an increase in *MT-1G* and *MT-2A* expression was observed in both cell lines, whereas in ARO cells, an additional increase in *MT-1A*, *MT-1F*, *MT-1H*, and *MT-1X* has been noted. Moreover, in these cells, an increase in cellular proliferation was observed, with a shift of cells from the G0/G1 to the S-G2 phase of the cell cycle (Liu et al. 2007, 2009). However, the results of these studies, showing an increased cell proliferation in relation to an increase in the expression of some MT isoforms, should be interpreted with caution, as the study design does not allow a clear characterization of the role of these isoforms in the cell proliferation process, as shown by other authors in case of *MT-1G* (Liu et al. 2007, 2009; Ferrario et al. 2008; Fu et al. 2013).

As differential expression of MT isoforms has been observed between normal, noncancerous thyroid lesions and malignancies, it has been proposed that MTs may be used as a marker of malignancy of tumors originating from this organ (Finn et al. 2007). Moreover, the findings by Krolicka et al. have shown that MT-1/2 expression, which is significantly upregulated in FTC as compared to follicular adenoma, could be also used in the differential diagnosis of these two lesions when none of the histological features (such as the presence of cancer cells in blood vessels and thyroid capsule invasion) could be identified in routine diagnostic hematoxylin and eosin stained tumor sections (Krolicka et al. 2010).

4.7 Respiratory System

4.7.1 *Nasal Cavity and Pharynx*

MT-1/2 expression has been studied in both benign and malignant lesions of the nasopharynx. MT-1/2 expression in the oral mucosa has been shown to be confined to the cytoplasm of basal and parabasal cells. It has been shown that, in dysplastic lesions, suprabasal layers also acquire MT-1/2 expression (Sundelin et al. 1997; Johann et al. 2008; Pontes et al. 2009). MT-1/2 immunoreactivity has been positively correlated with the severity of the dysplasia and with Akt expression (Pontes et al. 2009). The highest levels of MT-1/2 have been noted in cells of invasive squamous cell carcinomas and nuclear–cytoplasmic expression of MT-1/2 has, likewise, been observed for the dysplastic oral mucosa (Szelachowska et al. 2008; Pontes et al. 2009).

The involvement of MT-1/2 expression in oral squamous cell cancer (OSCC) development has been confirmed by several in vivo studies using animal models (Liu et al. 2005; Fong et al. 2006). These studies point to the regulatory role of zinc on MT-1 expression and other markers related to carcinogenesis, e.g., cyclin-B2, carbonic anhydrase II, and keratin 14. Decreased zinc levels lead to an increase in the expression of these factors, though zinc dietary supplementation in rats leads to diminished expression of these carcinogenesis-related markers; a concomitant suppression of the cellular proliferation of oral mucosa could be observed (Liu et al. 2005). As shown for human oral mucosa, MT-1 expression increased with lesion progression in the hyperplasia–dysplasia–carcinoma experimental model (Fong et al. 2006).

The major etiological factors of oral squamous cell carcinoma (OSCC) development and progression are found to be areca (betel), nut chewing (mainly in Asia and South America), and cigarette smoking (Trivedy et al. 2002). The areca nuts contain high amount of copper and are capable of generating high levels of reactive oxygen species, which could cause damage to the oral mucosa and may induce MT-1/2 expression (Trivedy et al. 2002). Such mechanism of OSCC development is supported by findings of more frequent overexpression of the MT-1 isoform in cancers originating in long-term areca nut consumers (Lee et al. 2008). Furthermore, in vitro data confirm the clinical observations, as arecoline treatment of the GNM oral epithelial cell line originating from a patient with stage T2N2aM0 gingival carcinoma was associated with increased *MT-1* mRNA expression (Lee et al. 2008). These studies provide a possible explanation for MT-1/2 overexpression in oral squamous cell carcinomas.

The relationship between MT-1/2 expression in OSCC and the clinical and pathological characteristics of the patients has been intensely studied. In the majority of studies, MT-1/2 expression has been shown to have no impact on the proliferation of cancer cells, as measured by the number of Ki-67 or minichromosome maintenance protein-2 (MCM2) positive cells (Muramatsu et al. 2000; Cardoso et al. 2002; Szelachowska et al. 2008, 2009). Cancers characterized by positive nuclear MT-1/2 immunostaining have been shown to be associated with higher p53 expression (Cardoso et al. 2009). OSCC with nodal involvement present higher MT-1/2 expression levels than do cases without metastasis (Szelachowska et al. 2009). This observation could be explained by the increased invasiveness of cases with MT-1/2 expression, as a positive correlation between MT-1/2 immunoreactivity and that of laminin-5—an extracellular matrix protein regarded as an invasiveness marker—has been disclosed (Cardoso et al. 2009). The possible correlation between chemoresistance and MT expression has also been investigated. Human tongue cancer cell lines express the MT-1, MT-2, and MT-4 isoforms, but not MT-3. Treatment of these cell lines with cisplatin resulted in a significant increase in the expression of the MT-1 and MT-2 isoforms only in cisplatin-resistant cells, pointing to their potential role in chemoresistance mechanisms (Nakano et al. 2003). However, no differences in MT-1/2 expression levels between treated and untreated oral squamous cell carcinoma patients could be observed, minimizing the significance of the results obtained in vitro (Muramatsu et al. 2000; Nakano et al. 2003). Nevertheless, high expression of MT-1/2 has been shown to be associated with poor patient prognosis as a single marker (Cardoso et al. 2002; Szelachowska et al. 2008, 2009) or in combination with high p53 expression (Cardoso et al. 2009).

4.7.2   Larynx

MT-1 and MT-2 have been shown to be upregulated in laryngeal squamous cell cancers, as compared to noncancerous tissues (Pastuszewski et al. 2007). Furthermore, their expression increases with growing malignancy of the lesions, with the level being lowest in epithelial dysplasia, higher in carcinomas in situ, and highest in invasive carcinomas (Ioachim et al. 1999). MT-1/2 expression has been found to correlate positively with the expression of PCNA in both benign and malignant lesions (Ioachim et al. 1999); however, in the study of Pastuszewski et al., no correlation of MT-1/2 expression with Ki-67 antigen positivity in cancer cells was observed (Pastuszewski et al. 2007). The prognostic significance of MT-1/2 expression in laryngeal squamous cell cancers is still under discussion, as divergent results have been obtained. For example, high levels of expression of MT-1/2 have been shown by Brown et al. to be a poor prognostic factor (Brown et al. 2003), but such relationship was not reported by other authors (Pastuszewski et al. 2007).

The altered expression of *MT-2A* and *MT-1E* isoforms probably contributes to the development and progression of laryngeal squamous cell cancers (Tan et al. 2005). Low levels of these isoforms' mRNAs have been detected in TW01 and HEp-2 laryngeal cancer cell lines. The expression of these two isoforms was lower than in embryonic lung fibroblast cell line (MRC-5) and on similar level in three nasopharyngeal cancer cell lines: HK1 (well differentiated), TW01 (moderately differentiated), and CNE2 (poorly differentiated) (Tan et al. 2005).

4.7.3   Lungs

MTs have been studied in nonmalignant lung tissue, as well as in non-small cell lung carcinomas (NSCLC) and small-cell lung carcinomas (SCLC) (Fig. 4.4). In normal lung tissue, MT expression in epithelial cells seems to protect this organ from various damaging factors, including ozone (Inoue et al. 2008), nonessential metal ions (Kenaga et al. 1996; Jia et al. 2004b; Waalkes et al. 2005), and benzo[a] pyrene-induced carcinogenesis (Takaishi et al. 2010). However, chronically elevated level of MTs in epithelial cells may contribute to the later induction of the carcinogenesis process (Person et al. 2013). The peripheral lung epithelia cell line (HPL-1D), under the influence of low, noncytotoxic levels of cadmium, acquired after 20 weeks of incubation phenotypic changes resembling those of invasive carcinoma cells, such as increased MMP-2 activity and invasiveness as well as augmented colony formation in soft agar. Furthermore, these cells were characterized by increased proliferation rates and increased growth rates in serum-free media. Chronic cadmium exposure induced the expression of some oncoproteins (K-RAS, N-RAS, and cyclin-D1) and decreased the expression of tumor suppressor genes p16 and SLC38A3 (sodium-coupled neutral amino acid transporter 3) on the protein level (Person et al. 2013). Increased expression levels of MT-1A and

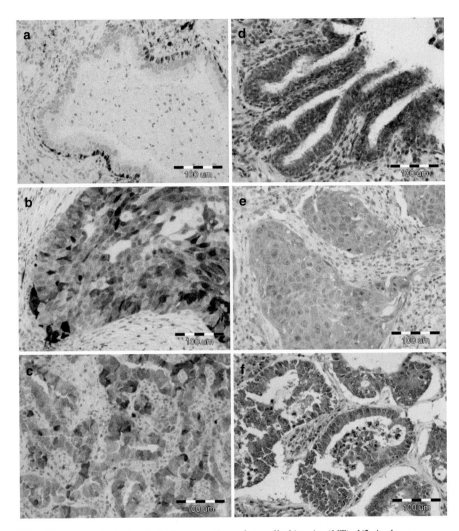

Fig. 4.4 *Immunohistochemical demonstration of metallothionein (MT) 1/2 in lung cancer.*
Expression of MT 1/2 in nuclei (**a**) and cytoplasm and nuclei (**d**) of typical respiratory epithelium
of nonmalignant lung tissue. Heterogeneous expression of MT 1/2 in cytoplasm and nuclei (**b**) and
cytoplasm (**e**) of squamous cell lung cancer as well as cytoplasm and nuclei (**c**) and cytoplasm (**f**)
of lung adenocarcinoma. Archival sections from the Department of Histology and Embryology,
Wroclaw Medical University, Wroclaw, Poland

MT-2A were also observed in cadmium-treated cells. Thus, it seems that MTs may
protect in vitro the cells from damaging factors up to the occurrence of some critical
event, but once the carcinogenetic switch turns on, they may contribute to tumor
progression (McGee et al. 2010; Person et al. 2013).

It has been shown that MT-1/2 expression increases in both types of NSCLC
(SQC, squamous cell carcinoma and AC, adenocarcinoma), whereas it decreases in

SLCL in comparison to the control nonmalignant lung tissue (Fig. 4.4) (Theocharis et al. 2002). In both types of NSCLC, MT-1/2 immunoreactivity has been noted in both cancer and stromal cells, though an association between its expression with clinical and pathological data has not been noted. It was found that mRNA expression levels of *MT-1B*, *MT-1F*, *MT-1G*, *MT-1H*, *MT-1X*, and *MT-3* were significantly upregulated, whereas that of *MT-1E* was significantly downregulated in NSCLC, as compared to control lung tissue (Werynska et al. 2013a, b). So far, no expression of MT-4 has been found in lung cancer or normal lung tissues (Werynska et al. 2013b). Interestingly, although MT-1/2 immunoreactivity clearly increased in cancer cells in comparison to normal pneumocytes, a translocation of the immunoreactivity from the nucleus to cytoplasm during NSCLC carcinogenesis has been observed in the case of MT-3 (Werynska et al. 2013a). Furthermore, although nuclear MT-3 expression could be also noted in cancer cells, the immunoreactivity of this isoform decreased with increasing malignancy grade of the SQC and AC cases analyzed (Werynska et al. 2013a).

A number of studies have investigated the relationship between MT-1/2 and MT-3 expression and the NSCLC patients' clinical and pathological data. The extent of MT-1/2 immunoreactivity has been found to correlate positively with proliferation markers (Ki-67 antigen and MCM-2), confirming its role in cancer cell proliferation (Werynska et al. 2011). Although high immunohistochemical signals of MT-1/2 predicted poor patient outcome (Dziegiel et al. 2004b), the recent study of Werynska et al. did not confirm the previous findings (Werynska et al. 2011). However, more detailed analysis of MT-1/2 isoforms on the mRNA level has revealed that increased expression of *MT-1F* and *MT-2A* predicted poor patient outcome (Werynska et al. 2013b). In regard to MT-3, its low mRNA and cytoplasmic expression in cancer cells has been shown to be correlated with poor patient outcome (Werynska et al. 2013a). The prognostic impact of MT-1/2 expression in SCLC has been evaluated to date only in a single study in which its high nuclear immunoreactivity has been shown to be an independent factor for short-term survival (Joseph et al. 2001).

4.8 Alimentary System

4.8.1 Salivary Glands

MTs have been shown to affect salivary gland development (Hecht et al. 2002). Laminin-1, a major component of the epithelial cell basement membrane, has been identified as inducing the expression of *MT-1B*, *MT-1F*, and *MT-2A* isoforms in human salivary gland (HSG) cells (Hecht et al. 2002). Further studies have revealed that overexpression of *MT-1F* does not alter the proliferation rate of HSG cells; however, these cells are characterized by augmented growth following the supplementation of low concentrations of zinc and copper ions in the medium, as compared to control cells (Hecht et al. 2002). Moreover, *MT-1F*-overexpressing cells are larger and have been shown to form acini and aggregate faster, presenting higher adhesive capabilities than the control-transfected parental cells. *MT-1F*-overexpressing cells, upon injection into nude mice, formed smaller and more differentiated tumors, confirming MT-1F's role in the differentiation of salivary gland cells (Hecht et al. 2002).

MT-1/2 expression has been studied in normal, benign, and malignant lesions of salivary glands originating from epithelial and myoepithelial cells (Sunardhi-Widyaputra et al. 1995; Gao et al. 1997; Hecht et al. 2002; Ogawa 2003; Alves et al. 2007). Although MT-1/2 expression was present in the epithelial and myoepithelial cells of these lesions, its exact role remains to be determined, as the abovementioned studies were performed on small groups of patients.

4.8.2 Esophagus

The expression of MTs has been studied in the normal mucosa of the esophagus and its precancerous lesions (Barrett's metaplasia and dysplasias of the squamous epithelia), as well as in both of its carcinoma types—esophageal squamous cell carcinoma (ESCC) and adenocarcinoma (EAC) (Fic et al. 2013). Altered expression of MT-1/2 and MT-3 has been found in cells of normal mucosa of the esophagus, Barrett's metaplasia (precancerous lesion of the EAC), and EAC (Coyle et al. 2002a; Li et al. 2003; Peng et al. 2011). It was found that MT-1 and MT-2 expression increases with the progression of histological changes from normal esophageal epithelium, through Barrett's esophagus to EAC (Li et al. 2003). Moreover, the intensity of MT-1/2 expression correlated negatively with the extent of apoptosis in the studied lesions, confirming the role of these isoforms in EAC carcinogenesis (Li et al. 2003). MT-1/2 expression was shown to be elevated in Barrett's epithelium as compared to cells of unchanged esophageal mucosa and EAC both of which presented comparable expression levels (Coyle et al. 2002a). An increase in MT-1/2 expression has also been observed in mouse models during progression of lingual and esophageal noncancerous lesions to invasive cancers (Fong et al. 2006). In contrast to MT-1/2, whose expression seems to increase with the extent of dysplastic changes in the esophagus, *MT-3* expression has been found to be downregulated in Barrett's metaplasia and EAC, as compared to the squamous epithelium of the esophagus and normal glandular gastric mucosa (Peng et al. 2011). *MT-3* downregulation in Barrett's metaplasia and EAC seems to result from hypermethylation of the *MT-3* gene promoter, as a correlation has been observed between the methylation intensity of specific CpG nucleotides and MT-3 expression (Peng et al. 2011). Furthermore, the promoter of the *MT-3* gene was also hypermethylated in the FLO-1 and OE33 EAC cell lines. These changes were also associated with repressive histone modifications, confirming the epigenetic regulation of *MT-3* expression during the pathogenesis of EAC (Peng et al. 2011). It seems that the extent of *MT-3* methylation may be of prognostic significance in EAC, as tumors with hypermethylation in the *MT-3* promoter were characterized by the presence of lymph node metastases and by advanced clinical stage (Peng et al. 2011). As in EAC, the *MT-3* promoter has also been found to be hypermethylated in ESCC, although no association with patients' clinical or pathological data could be noted (Smith et al. 2005). Hypermethylation has also been noted in the promoter region of *MT-1G* in ESCC (62 % of the cases), in comparison to healthy esophageal mucosa and low- and high-grade dysplasia (9 %, 12 %, and 57 %, respectively) (Roth et al. 2006). Similar observations have been noted in esophageal mucosa cytology specimens, where high-grade dysplastic lesions and ESCC presented with the highest levels of *MT-1G* methylation (Adams et al. 2008). In addition, Oka et al. have observed the hypermethylation of *MT-1M* in ESCC as compared to noncancerous esophageal mucosa (Oka et al. 2009).

The involvement of MT-1/2 in the pathogenesis of esophageal cancer seems to be mediated by its indirect interaction with the *ECRG2* (Esophageal Cancer Related Gene 2) gene, which functions as an oncosuppressor. Co-expression of ECRG2 has been shown to occur with *MT-1G, MT-1H,* and *MT-2A* in ESCC cell nuclei (Cui et al. 2003). When both *MT-2A* and *ECRG2* were overexpressed in esophageal cancer EC103 and hepatocellular carcinoma Bel7402 cell lines, the cell proliferation rates decreased. However, when both protein and *ECRG2* expression were inhibited in the cell lines studied, an increase in cellular proliferation rate has been noted (Cui et al. 2003). A divergent effect has been observed following the overexpression of *MT-3* in human esophageal carcinoma Eca-109 and TE13 cell lines (Tian et al. 2013). These cell lines with the *MT-3* gene transfection are characterized by decreased proliferation rates and higher proportions of cells in the G0/G1 phase, as well as fewer cells in the S phase of the cell cycle. Also, a significant increase in apoptosis rate has been observed in these cells, as compared to the blank vector-treated controls (Tian et al. 2013). The results of these studies suggest that particular MT isoforms may play differential roles in the pathogenesis of EAC and ESCC. In line with their known functions, the majority of the overexpressed *MT-1* and *MT-2* isoforms seem to be associated with the progression of dysplastic lesions in the esophagus, whereas *MT-3, MT-1G,* and *MT-1M* may exert a protective effect on the esophageal epithelium, preventing tumor development (Cui et al. 2003; Li et al. 2003; Smith et al. 2005; Roth et al. 2006; Adams et al. 2008; Oka et al. 2009; Peng et al. 2011). However, further studies are needed to elucidate the role of particular MT isoforms in the pathogenesis of EAC and ESCC.

MT-1/2 expression has also been studied in regard to its prognostic value in the esophageal cancer. Hishikawa et al. used immunohistochemical methods and found the expression of MT-1/2 in cytoplasm and nuclei of ESCC cancer cells. High MT-1/2 expression has been significantly associated with the presence of distant metastases and poor patient outcome (Hishikawa et al. 1999). Moreover, MT-1/2 expression correlates positively with the expression of PCNA, confirming the role of these isoforms in the regulation of cell proliferation. The expression of *MT-1A* and *MT-2A* mRNA in ESCC cells has been confirmed by in situ hybridization and found to be associated with the more frequent presence of metastases in lymph nodes and distant organs (Hishikawa et al. 1999). MT-1/2 expression has also been found to be a useful potential predictive factor for cisplatin-based chemotherapy in a cohort of 68 ESCC patients (Hishikawa et al. 1997). Similar results have also been demonstrated in a group of 36 ESCC patients undergoing treatment with a combination of chemotherapy (5-fluorouracil and cisplatin) and radiotherapy (Sunada et al. 2005). In that study, MT-1/2 expression and negative expression of p53 and CDC25B (M-phase Inducer Phosphatase 2) proteins were found to be independent predictive factors in the analyzed patients (Sunada et al. 2005). Furthermore, MT-1/2 expression has also been recognized as a potential responsiveness marker to neoadjuvant chemoradiotherapy (CRT) therapy in a group of 30 ESCC patients (Yamamoto et al. 1999). However, results of the study of Nomiya et al. suggest that MT-1/2 expression is not inducible in cancer cells of ESCC following radiotherapy—indicating its limited role as a potential predictive factor for preoperative radiotherapy (Nomiya et al. 2004). Thus, it may be stated that MT-1/2 expression in ESCC is a better predictive factor for chemotherapy than for radiotherapy (Hishikawa et al. 1997, 1999; Yamamoto et al. 1999; Nomiya et al. 2004).

4.8.3 Stomach

Gastric cancer is one of the most commonly diagnosed malignancies in the world. Currently, multiple predisposing environmental and biological factors have been recognized. *Helicobacter pylori* is regarded as a class A carcinogen; its infection causes chronic inflammation, which subsequently leads to the atrophy of the mucosa, and its metaplastic and dysplastic transformation, which will, over time, result in the development of the intestinal type of gastric cancer, according to the Lauren classification (Panani 2008). Lines of evidence suggest that the MT-1/2 isoform may function as a protective agent which suppresses the carcinogenic effects of *Helicobacter pylori* infection. For example, it has been shown that mice lacking MT-1/2 were more susceptible to colonization and gastric inflammation caused by these bacteria (Tran et al. 2003). The protective effect of MT-1/2 molecules is probably based on their scavenging effects toward ROS, since short-term zinc treatment protected mice against *H. pylori*-induced gastric inflammation (Tran et al. 2005). This observation suggests that zinc—a potent MT-1/2 inductor—may be effective in attenuating gastric mucosa inflammations and that this effect could be attributed to the increased expression of MT-1/2 in gastric mucosa. The postulated protective role of MT-1/2 in gastric inflammation has been supported by experiments which showed that mice lacking its expression tended to have worse disease course (Mita et al. 2008, 2012). The majority of independent clinical studies performed on human tissue samples have shown that MT-1/2 expression was downregulated in gastric cancer specimens and *H. pylori*-infected mucosa, as compared with corresponding noncancerous mucosa (Janssen et al. 2000; Tuccari et al. 2000; Mannick et al. 2004; Mitani et al. 2008). The immunohistochemical analysis of the expression of MT-1/2 in gastric mucosa in relation to the presence of

H. pylori infections and early stage gastric cancers revealed that infected gastric mucosa was characterized by lower levels of MT-1/2 expression than the noninfected one (Mitani et al. 2008). Moreover, intensity of MT-1/2 immunoreactivity in the mucosa was not affected by the presence of gastric cancer in adjacent tissue fragments. Moreover, it was found that the expression of MT-1/2 isoforms increased and reached levels comparable to those noted in unaffected mucosa following eradicative therapy with antibiotics (Mitani et al. 2008). Based on the results, the protective effect of MT-1/2 expression on the gastric mucosa may be also exerted through the shift in the number of proliferative and apoptotic cells, since high MT-1/2 expression was accompanied by a low number of Ki-67-stained mucosa cells and a high number of apoptotic cells (Mitani et al. 2008).

The recent studies of Pan et al. have identified the MT-2A isoform as a potent oncosuppressor that prevents the development of gastric cancer (Pan et al. 2013a, b). A decreased expression of MT-2A was presented on mRNA and protein levels in gastric cancer specimens, as compared to noncancerous gastric mucosa. Additionally, the expression of *MT-2A* mRNA was lower in gastric cancer cell lines (PAMC82, MKN45, SNU-5, SNU-16, and N87) than in GES-1 cells derived from normal gastric epithelium. The lowest *MT-2A* mRNA levels were noted in poorly differentiated or metastatic gastric cancer cell lines, namely BGC823, SGC7901, MGC803, AGS, SNU-1, RF-1, and RF-48. The oncosuppressory function of MT-2A has been ascribed to its inhibitory role on NF-κB pathway activation (Pan et al. 2013a). Furthermore, *MT-2A* gene has been shown to inhibit the growth of gastric cancer cell lines (BGC823, SGC7901, and AGS) through apoptosis and G2/M arrest, which negatively regulates the NF-κB pathway through upregulation of IκB-α and downregulation of p-IκB-α and cyclin D1 expression (Pan et al. 2013a). Interestingly, MiRNA-23a has been identified as a suppressor of *MT-2A* expression, indicating MT-2A involvement in gastric cancer development and progression (An et al. 2013).

MT-3 has been the least studied isoform of MT in gastric mucosa. Similarly to MT-1/2, decreased *MT-3* expression was observed in gastric cancer tissues, compared with noncancerous gastric mucosa (Ebert et al. 2000; Deng et al. 2003a, b). Hypermethylation of the *MT-3* intron1 CpG island has been identified as responsible for the underexpression of this isoform in gastric cancer tissues and cell lines (Deng et al. 2003a, b).

Only two studies have shown MT-1/2 expression to be higher in gastric cancer than in the corresponding normal mucosa (Ebert et al. 2000; Galizia et al. 2006). In the noncancerous gastric tissues, MT-1/2 immunoreactivity was restricted to the superficial gastric epithelium toward the top of the gastric glands and the foveolar neck of the gastric glands. Areas of dysplasia and intestinal metaplasia, as well as gastric cancer cells, were characterized by intense MT-1/2 expression, which has been confirmed using Northern blot analysis, showing an overexpression of *MT-2A* in gastric cancer (Ebert et al. 2000).

The prognostic and predictive impact of MT expression has also been addressed in gastric cancer cell line models and clinical samples. It has been shown that in gastric cancer, tumor stage and type, as well as differentiation grade, are independent of MT-1/2 immunoreactivity (Ebert et al. 2000). In another study of gastric cancer, high immunoreactivity of MT-1/2 expression was shown to be a poor prognosis factor, indicating its role in tumor progression (Galizia et al. 2006). In line with this observation are the results related to the chemoresistance of gastric cancer cells. It has been found that high expression of MT-1/2 in gastric cancer cells is associated with resistance to irinotecan (Chun et al. 2004) and cisplatin (Choi et al. 2004). The latter may be overcome by treating cells with heptaplatin, which has been shown to be more effective than cisplatin in gastric cancer cells with high expression levels of MT-1/2 (Choi et al. 2004). However, in clinical studies low *MT-2A* expression was found to be a marker of poor prognosis in the analyzed patient groups (Pan et al. 2013a, b). Taking into account the current results, further research is needed to identify the roles of particular MT isoforms in gastric cancer.

MT-1/2 expression has also been evaluated in gastrointestinal stromal tumors (GIST) where, in comparison to gastric cancers, lower *MT-2A* mRNA expression was found (Soo et al. 2011). High immunoreactivity levels of MT-1/2 in GIST have been shown to be associated with worse prognosis (Perez-Gutierrez et al. 2007; Pula et al. 2013).

4.8.4 Intestine

As in the case of the gastric mucosa, the expression of MT-1/2 isoforms in the colorectal mucosa seems to function as a potent oncosuppressor of carcinogenesis and to protect against damage (Fig. 4.5) (Pedersen et al. 2009; Tran et al. 2009; Fic et al. 2013). In a model of methotrexate-induced intestinal inflammation, MT-null mice suffered more severe histological damage and presented with augmented neutrophil infiltration (Tran et al. 2009). Furthermore, decreases in MT-1/2 immunohistochemical reactions have been seen in colorectal adenomas and adenocarcinomas, as compared to normal noncancerous mucosa (Janssen et al. 2000, 2002; Yan et al. 2012). Recent studies documented the expression of particular isoforms in colorectal cancer development. A decrease in *MT-1F*, *MT-1G*, *MT-1X*, and *MT-2A* mRNA has been noted in microdissected cells of colorectal adenocarcinomas, as compared to the corresponding glandular cells of normal mucosa (Yan et al. 2012). Low *MT-1F* expression has also been observed in the HCT116, HT29, SW620, CW-2, and LoVo colorectal cancer cell lines (Yan et al. 2012). Exogenous *MT-1F* expression in the RKO cell line increased apoptosis and inhibited cell migration, invasion, and adhesion as well as the cells' in vivo tumorigenicity in comparison to the control cells (Yan et al. 2012). The observed downregulation of the *MT-1F* gene in the majority of the colon tumor tissues studied was mainly due to loss of heterozygosity, while the CpG island methylation of the *MT-1F* gene promoter region has only been observed in poorly differentiated, microsatellite instable RKO and LoVo cell lines (Yan et al. 2012). An oncosuppressory role of particular MT isoforms has also been suggested by findings of decreased mRNA levels of *MT-1E*, *MT-1F*, *MT-1G*, *MT-1H*, *MT-1M*, *MT-1X*, and *MT-2A* in colorectal cancers, compared with normal healthy mucosa (Arriaga et al. 2012b). A CpG

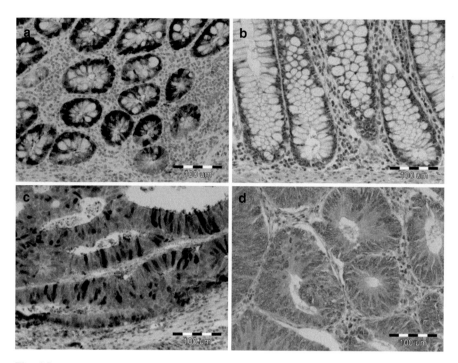

Fig. 4.5 *Immunohistochemical demonstration of metallothionein 1/2 in normal human large intestine* (**a**, **b**) *and colon adenocarcinoma* (**c**, **d**). Immunoreactivity was present in cytoplasm and nuclei (**a**, **c**) or only in cytoplasm (**b**, **d**). Archival sections from the Department of Histology and Embryology, Wroclaw Medical University, Wroclaw, Poland

island methylation in the promoter of the *MT-1G* gene has been shown to be responsible for its low expression in adenocarcinomas (Arriaga et al. 2012b). Further studies have revealed that overexpression of the *MT-1G* isoform sensitizes colorectal cell lines to the chemotherapeutic agents oxaliplatin (OXA) and 5-fluorouracil (5-FU). Overexpression of *MT-1G* also leads to enhancement of p53 and inhibition of NF-κB activity, which might partially explain the increase in the susceptibility of colorectal cancer cells to these agents (Arriaga et al. 2014). In a small patient group, a motif of thymine repeats (T20) within the 3′untranslated region of the *MT-1X* gene (MT-1XT20) showed a monomorphic pattern in normal tissue samples and a frequency of 100 % in cancer cases presenting high microsatellite instability (MSI-H) (Giacomini et al. 2005). Further testing in 340 colorectal specimens revealed its high sensitivity (97.1 %) and specificity (100 %) in detecting MSI-H patients (Morandi et al. 2012). The analysis of MT-1XT20 might be a good addition to the current markers of MSI assessment, as it showed higher sensitivity and specificity in the abovementioned patient cohort than the other markers tested (Morandi et al. 2012).

The prognostic impact of MT-1/2 expression has been frequently analyzed and has yielded discrepant results. Some studies identify low MT-1/2 immunoreactivity as a poor prognostic factor (Schmitz et al. 2009; Arriaga et al. 2012a, b), whereas others have shown that high MT-1/2 expression is correlated with shorter patient survival (Hishikawa et al. 2001; Janssen et al. 2002). Some reports have shown that MT-1/2 expression, though correlated positively with Ki-67 antigen expression, has no impact on patient prognosis (Dziegiel et al. 2003, 2004a).

4.8.5 Liver

Isoforms of the MT-1/2 family have been shown to be involved in hepatic carcinogenesis. The expression of these isoforms has also been studied in benign hyperplastic and malignant lesions of this organ, in which oxidative stress, induced as result of HBV or HCV infections or chronic alcohol intake, contributes strongly to the carcinogenic process (Block et al. 2003; Varela et al. 2003). It has been shown in mice that MT-1/2 expression suppresses hepatic hyperplasia induced by the HBV infection (Quaife et al. 1999), whereas MT-null mice were characterized by a higher susceptibility to the development of hepatocellular carcinoma upon cisplatin treatment at clinically relevant doses or diethylnitrosamine intoxication (Waalkes et al. 2006; Datta et al. 2007; Majumder et al. 2010). Lines of evidence suggest that, during carcinogenesis, hepatocytes lose MT-1/2 expression, as this downregulation has been observed in cancer cells of hepatocellular carcinoma, compared with the surrounding hepatocytes using immunohistochemical methods (Endo et al. 2004; Datta et al. 2007; Tao et al. 2007). Molecular studies confirm the particular downregulation of MT-1/2 isoforms on the mRNA level. For example, Datta et al. have shown decreases of the mRNA levels of *MT-1A, MT-1E, MT-1G, MT-1H, MT-1X*, and *MT-2A* (Datta et al. 2007), whereas Mao et al. showed decreased levels of *MT-1A, MT-1F, MT-1G, MT-1H*, and *MT-1M* (Mao et al. 2012). These effects were ascribed to the hypermethylation of the CpG islands in *MT* gene promoters (Datta et al. 2007; Mao et al. 2012). In the latter study, however, no differences were seen in regard to *MT-1B, MT-1E*, or *MT-1X* expression in normal liver or hepatocellular carcinoma samples (Mao et al. 2012). In line with this, correspondingly significantly lower MT-1/2 concentrations in the serum of patients with hepatocellular carcinoma have been noted, in comparison to patients with noncancerous liver condition (Nakayama et al. 2002).

The proposed oncosuppressory role of MT-1/2 isoforms has also been confirmed using cell-based experiments (Datta et al. 2007; Mao et al. 2012). Datta et al., in addition to the observed downregulation of several MT-1/2 isoform expressions in malignant lesions, did not find changes in the expression levels of *MTF-1* or zinc transporter-1 (*Zn-T1*) (Datta et al. 2007). The alterations in MT-1/2 transcriptional activity leading to their decreased expression in malignant lesions have been identified as being caused by alterations in the phosphatidylinositol 3-kinase (PI3K)/Akt signaling pathway, which has been shown to be frequently upregulated in human cancers (Vivanco and Sawyers 2002). The inhibitors of PI3K and Akt increased the expression of MT genes in hepatocellular carcinoma cells, but not in hepatocytes (Vivanco and Sawyers 2002). The repression of MT-1/2 expression was due to the activation of glycogen synthase kinase-3 (GSK-3), a downstream target of the PI3K/Akt pathway. Upon activation, this factor in turn decreases the levels of active CCAAT/enhancer binding protein alpha (C/EBP alpha), which leads to the diminished MT-1/2 expression levels exclusively found in hepatocellular carcinoma cells (Datta et al. 2007). It has been also shown that the DNA-binding activity of C/EBP alpha and its phosphorylation at T222 and T226 by GSK-3 are required for MT expression (Datta et al. 2007). The loss of the expression of the MT-1/2 isoforms during hepatic carcinogenesis has also been associated with the activation of NF-kB target genes (Majumder et al. 2010). Upon diethylnitrosamine intoxication in wild-type and MT-null mice, increased NF-κB activity in the liver nuclear extracts of both genotypes has been observed; however, different protein complexes developed. In the MT-null mice, the p50/65 heterodimer (transcriptional activator) complex was predominantly observed, as opposed to the p50 homodimer (transcriptional repressor) seen in wild-type mice (Majumder et al. 2010). Furthermore, increased expression of Pai-1, c-Jun, c-Fos, c-Myc, Ets2, and ATF3 expression in the MT-null mice has been noted (Majumder et al. 2010). The *MT-1M* isoform has been identified as a potent oncosuppressor for hepatocellular carcinoma (Mao et al. 2012). The introduction of this isoform to hepatocellular carcinoma Hep3B cell line, which presents no basal expression of MT-1M, suppressed cell growth in vitro and in vivo and sensitized the cells to TNFα-induced apoptosis (Mao et al. 2012).

Analysis of patients' clinical and pathological data in regard to MT-1/2 expression revealed that the expression levels assessed by IHC in cancer cells of hepatocellular carcinoma are inversely correlated with the malignancy grade of the tumors and the clinical advancement stage (Endo et al. 2004; Tao et al. 2007). Furthermore, high MT-1/2 expression has been shown to be a good prognostic factor in individual studies and in a recently published meta-analysis (Tao et al. 2007; Gumulec et al. 2014). The downregulation of *MT-1G* has also been observed in hepatoblastoma, where its decreased mRNA levels predicted poor patient outcome (Nagata et al. 2003; Sakamoto et al. 2010).

4.8.6   Pancreas

Metallothionein-1/2 expression has been studied in both the endocrine and the exocrine cells of the pancreas. MT-1/2 expression has been identified in acini and in four major types of endocrine islet cells, as well as intralobular ducts; however, no expression has been noted in the large pancreatic ducts (Ohshio et al. 1996; Tomita 2000; Sliwinska-Mosson et al. 2009). Decreased MT-1/2 expression has also been observed in endocrine neoplasms of the pancreas, in comparison to normal islet cells; however, the prognostic significance of this finding remains to be clarified (Tomita 2000, 2002; Tomita and Matsubara 2000). Decreased MT-1/2 immunoreactivity has been found in serous cystadenomas, as compared to the normal exocrine cells of the pancreas (Sliwinska-Mosson et al. 2009). However, MT-1/2 immunohistochemical reaction was increased in pancreatic cancer, as compared to non-transformed pancreatic tissues (Ohshio et al. 1996; Sliwinska-Mosson et al. 2009). The analysis of MT-1/2 immunoreactivity in 75 cases of pancreatic cancer in relation to clinical and pathological characteristics identified the immunohistochemical reaction of these isoforms as a poor prognostic indicator (Ohshio et al. 1996). Moreover, MT-1/2 expression in cancer cells has also been associated with the presence of lymph node metastasis and higher malignancy grades of the analyzed cancers (Ohshio et al. 1996). Although this study showed a negative impact on patient prognosis, MT-1/2 expression was not related to cisplatin resistance of pancreatic cancer cell lines (Ohshio et al. 1996).

4.9   Urinary Tract

4.9.1   Kidney

It has been observed that, in normal kidneys, MT-1/2 expression occurs only in the cytoplasm of tubular epithelial cells, whereas in renal neoplasms, a patchy pattern is found with variable subcellular staining—i.e., cytoplasmic, nuclear, and membranous staining (Tüzel et al. 2001; Ishii et al. 2001; Cherian et al. 2003). Tissues adjacent to the tumor revealed stronger MT-1/2 expression than cancer lesions (Ishii et al. 2001). Approximately 80 % of renal cell carcinomas (RCC) develop from the epithelium of the proximal tubuli (Cerulli et al. 2006). The intensity of MT-1/2 expression decreases in high-grade cancers, indicating the role of MT-1/2 in processes of tumor growth and differentiation (Saika et al. 1992; Izawa et al. 1998; Mitropoulos et al. 2005). The extent of MT-1/2 expression is inversely correlated with survival and staging, which is possibly associated with rapid tumor growth (Izawa et al. 1998; Mitropoulos et al. 2005). However, in another study only the relationships of MT-1/2 expression with survival and grading were confirmed (Tüzel et al. 2001). Additionally, an association between decreased MT-1/2 expression and increased apoptosis, which is characteristic of cancers with invasive growth patterns, was found (Zhang et al. 1998).

The expression of MT-3 has been assessed in normal tissue and neoplastic kidney lesions. Hoey et al. obtained results that are in line with the evidence for the specific functions of MT-3. The increase of *MT-3* mRNA can be considered as a response to metal exposure, but in renal cancer cases changes were not found (Hoey et al. 1997). Interestingly, in view of the histological type of neoplasm, significantly higher IHC expression of MT-1/2 has been found in sarcomatoid tumors, in relation to clear cell, papillary, granular, and chromophobe tumors (Tüzel et al. 2001). On the molecular level, the expression of MT isoforms has been assessed in a few studies by RT-PCR in normal and neoplastically altered renal tissues. The transcript levels of *MT-2A* were significantly upregulated, whereas *MT-1A* and *MT-1G* were downregulated in renal cancer (Nguyen et al. 2000). Considering that the methylation profile of the *MT* gene might be altered in renal cancer, studies of hypermethylation also have been performed. Dalgin et al. found hypermethylation of *MT-1G* in cancer cases, which partially conforms to the observations of Nguyen et al. and Alkamal et al. (Nguyen et al. 2000; Dalgin et al. 2008; Alkamal et al. 2015). The detection of the hypermethylation of highly downregulated genes may become a valuable tool in the diagnostics of renal cancer assessment.

Being cysteine-rich proteins, MTs are responsible for detoxification processes, mostly related to heavy metal ions sequestration and inactivation of reactive oxygen species. It is well known that tumors localized in the urinary system are closely related to environmental factor intoxication (Cerulli et al. 2006). Studies conducted on animal models with MT-knockout mice have confirmed that the absence of MT-1/2 expression results in increased Pb-induced cancerogenesis, which directly shows the protective function of MT-1/2 on renal tissue (Waalkes et al. 2004; Tokar et al. 2010). However, MTs belong to the main factors responsible for chemoresistance phenomena. Studies on mice models showed that induced MT-1/2 expression by Zn exposure simultaneously induced chemoresistance to cisplatin (Kennette et al. 2005). This might result from the activation of telomerase (one of the factors responsible for unlimited cell division), which was observed in Zn-exposed cells and MT-1/2-overexpressing cells (Nemoto et al. 2000).

4.9.2   Urinary Bladder

The most abundant malignant neoplastic lesions originating from the urinary bladder are urothelial carcinomas—predominantly infiltrating urothelial or transitional cell carcinoma (TCC). Their principal cause in most countries seems to be tobacco smoking and, consequently, the influence of free radicals and heavy ions on the uroepithelium (Masson-Lecomte et al. 2014). It has been shown that the expression of MT-1/2 is increased in bladder tumors and is positively correlated with the grade of histological malignancy and the stage of clinical advancement (Saika et al. 1992; Ioachim et al. 2001; Saga et al. 2002). Zhou et al. have analyzed by IHC a broad group of TCCs and found no expression of MT-1/2 in benign lesions and low-grade cancers, a low incidence of expression in dysplastic lesions and high-grade cancers with no evidence of muscle invasion, and a significantly increased incidence of MT-1/2 expression in high-grade cancers with matrix invasion. The authors concluded that MT-1/2 may be a prognostic marker for cancer invasiveness into the underlying tissues of the bladder wall (Zhou et al. 2006a, b). Additionally, the observation of a positive correlation between MT-1/2 and proliferation marker PCNA expression and a negative correlation between MT expression and apoptosis supports the well-known antiapoptotic and pro-proliferative role of MTs in cell (Ioachim et al. 2001).

Increased MT-1/2 expression in TCCs is associated with lower overall survival, disease-specific survival, disease-free survival, and disease-free progression (Yamasaki et al. 2006). In view of this, it was even postulated to use MT-1/2 as a negative prognostic marker (Yamasaki et al. 2006). These findings have been confirmed by Hinkel et al., who showed that the elevation of MT-1/2 correlated with a decreased 5-year survival rate and increased 5-year recurrence and progression rates (Hinkel et al. 2008). Lynn et al. even suggested that elevated MT-1/2 might be a predictive marker of recurrence (Lynn et al. 2003). Moreover, MT-1/2 overexpression correlated with poorer survival and resistance to alkylating agents. These findings strongly support the protective role of MT during DNA damage (Jia et al. 2004a, b). Wulfing et al. stated also that "MT-positive" patients undergoing chemotherapy showed worse prognosis (Wülfing et al. 2007). Additionally, elevations of MT levels in tissues and fluids can be observed as a reaction to heavy metal ions or to factors that generate reactive oxygen species (i.e., chemotherapeutics) as a mechanism of detoxification. MT-Cd complexes were found to be increased in the urine of patients suffering from TCC (Wolf et al. 2009). Conversely, in studies with MT-knockout mice, N-butyl-N-(4hydroxybutyl)nitrosamine-induced tumors occurred much more often than in MT-wild-type mice (Kondo et al. 1999; Takaba et al. 2000). On the molecular level, particular MT isoforms are responsible for the processes of chemoresistance. For example, it was shown by HPLC and mass spectrometry that the MT-2A isoform was responsible for the sequestration of the drugs in human bladder tumor T24 cells (He et al. 2000). Similar observations have been made with RT 112 and CP3 cancer cell lines (Siegsmund et al. 1999). They postulated that decreased levels of MT-2A were mainly responsible for the chemoresistance (Siegsmund et al. 1999). Since then, many experiments have been conducted showing induced chemoresistance, mostly against cisplatin, adriamycin, chlorambucil, and melphalan associated with the overexpression of MT-1/2 in genetically modified mice (Saika et al. 1994; Satoh et al. 1994; He et al. 2000; Saga et al. 2004). There are also some studies which suggest possible suppression of the chemoresistance induced by MTs by using, e.g., propargylglycine (PPG) (Satoh et al. 1994). It has been suggested that cisplatin resistance may be overcome by the use of PPG, due to its additional nephroprotective activity (Satoh et al. 1994). Nevertheless, so far the only way to decrease toxicity and reduce or avoid chemoresistance is to administer cisplatin as a single, unfractionated injection, which maintains its antitumor activity with minimal MT-1/2 induction (Kondo et al. 2003). This observation seems to be important in view of another study, in which patients with undetectable expression of MT-1/2 responded completely to chemotherapy (Bahnson et al. 1994).

Molecular analyses also pointed toward other possible functions of MT isoforms. Wu et al. suggested that MT-1E could be necessary for cancer cell migration (Wu et al. 2008), while others found MT-2A and MT-1X to be present in the normal urinary bladder and found MT-1X overexpression only in cancer cases (Somji et al. 2001). It was postulated that the overexpression of MT-1/2 in bladder cancer may be a result of overexpression of the *MT-1X* gene (Somji et al. 2001). The expression of MT-3 mRNA and protein was also studied in bladder cancers. Sens et al. found that MT-3 was overexpressed in bladder cancer, and thus correlated positively with tumor grade, while in normal tissue, its expression was very weak (Sens et al. 2000). A discrepancy between protein and mRNA expression was shown, as the protein was found in the cells of high-grade cancers, while mRNA was not involved in malignant transformation, suggesting that a posttranslational mechanism is responsible for the presence of MT-3 in bladder cancer cases (Zhou et al. 2006a, b).

4.10 Organs of the Male Reproductive Tract

4.10.1 Testis

Weak expression of MT-1/2 has been observed in normal testicular tissue (i.e., mostly seminiferous tubuli) and benign lesions (seminoma), whereas strong reactions were demonstrated in malignant nonseminomas, mostly in the peripheral regions of the tumor (Chin et al. 1993; Cherian et al. 2003). Elevated expression of MT-1/2 is associated with tumor cell proliferation, aggressiveness, and drug resistance, whereas, surprisingly, poor prognosis correlates with decreases in its levels (Chin et al. 1993; Pedersen et al. 2009).

Considering the detoxifying function of MT-1/2, its expression in testicular cancer has been investigated. The poor synthesis of MT-1/2 in seminiferous tubules may predispose to Pb-induced cancerogenesis (Waalkes et al. 2004). On the other hand, it has been shown that prolonged Cd exposure induces abundant *MT* gene expression and leads to increased proliferation and progression of testicular cancer. These observations are supported by a study which showed an enhanced sensitivity to Pb intoxication in MT-knockout mice, which led to increased incidence of genitourinary neoplasms, including renal, testicular, and urinary bladder tumors (Tokar et al. 2010).

From a clinical point of view, analyses of the involvement of MTs in the resistance to some commonly used anticancer drugs seem to be important (Eid et al. 1998; Meijer et al. 2000; Mayer et al. 2003). For example, it was found that high MT-1/2 levels in testicular tumors predict a better response to chemotherapy, whereas tumors lacking MT-1/2 or demonstrating low MT-1/2 expression have worse predictions. These data do not support the hypothesis that MT overexpression could play a significant role in drug resistance, at least in this tumor type (Eid et al. 1998). Cisplatin, being a highly effective drug in anticancer therapy, has been investigated in many studies. Interestingly, the overexpression of MT-1/2 was found in germinal cell tumor cell lines, as well as in human tumors—despite the fact that overexpression of MT-1/2 has been observed in cisplatin-sensitive prostate tumors (Meijer et al. 2000). The studies of death-associated protein serine/threonine kinase 17A (STK17A) in murine testicular cancer cell lines showed that, during chemotherapy, MT-1/2 expression was induced by cisplatin, which resulted in growth suppression and apoptosis (Mao et al. 2011). The knockdown of STK17A in murine testicular cell lines massively decreased levels of cellular ROS and resulted in the resistance of cells to cisplatin. Subsequently, the upregulation of detoxifying and antioxidant genes, including *MT-1H, MT-1M*, and *MT-1X*, was observed. It has been suggested that induced MT expression in experimental testicular cancer cell lines could be a form of response to induced chemoresistance (Mao et al. 2011). Moreover, in view of the results of Meijer et al., who pointed out that MT-protein expression in primary germ-cell tumors did not differ between responding and nonresponding chemotherapy patients, it seems that MTs cannot be used as predictive markers in testicular cancer (Meijer et al. 2000).

4.10.2 Prostate Gland

The expression pattern of MTs in prostate cancer is still equivocal. There is evidence that, in normal prostate tissue, a differential expression of MT-1/2 is observed, ranging from intense expression in low-grade cancers to no expression in high-grade cancers (Moussa et al. 1997; Wei et al. 2008; Lee et al. 2009). El Sharkawy et al. have found MT-1/2 expression in both normal tissues and benign lesions, but contrary to the previously mentioned studies, they also described an evident increase in MT-1/2 expression with the grade of histological malignancy (El Sharkawy et al. 2006). Moreover, a positive correlation between staging, proliferation ratio, recurrence, and elevated MT-1/2 expression was shown (Athanassiadou et al. 2007). With increasing aggressiveness of the lesions, MT-1/2 immunoreactivity was shown to change from a nuclear localization to the cytoplasm. In normal prostate tissue, stronger expression was revealed in the peripheral organ region than in the central zone (Suzuki et al. 1991). Moreover, stromal and epithelial expression of MT-3 was demonstrated at low levels in normal tissue, while its intensity increased with the malignancy of prostate lesions (Garrett et al. 1999). Further studies revealed that MT-3 overexpression, analogically to MT-1F, can influence cancer growth and chemoresistance (Dutta et al. 2002). This is in line with the results of studies on cell lines pointing to the role of MT-3 overexpression in the processes of proliferation, invasion, and cancerogenesis (Juang et al. 2013). Some differences in the expression of MTs on the mRNA and protein levels were also reported. For example, in prostate cancer an increase in *MT-1A* mRNA level and a simultaneous decrease in the MT-1/2 proteins were found. It can partially be explained by posttranslational processing (Gumulec et al. 2012).

On the molecular level, the expression of MTs has been assessed in the normal prostate cell line RWPE-1 and in normal prostate tissue (Albrecht et al. 2008). Similar gene expression profiles (*MT-1E, MT-1X, MT-2A*) and analogous MT-1/2 protein expression were observed in the groups studied supporting the notion that the RWPE-1 cell line could be used as a reference cell line (Albrecht et al. 2008).

Moreover, a similar gene profile for normal tissue has been observed in previous studies, while in cancers, the downregulation of some genes—such as of *MT-1X*—was found (Garrett et al. 2000). It has been shown that SNPs (single nucleotide polymorphisms) of *MT-2A* were strongly associated with a decrease in or a lack of function of the encoded protein, resulting in the accumulation of Zn, Cd, Pb, and Cu ions in neoplastically changed prostate gland (Krześlak et al. 2013). SNPs represent analogous mechanism of genetic control to the hypermethylation phenomenon which was also observed in prostate cancer. It has been demonstrated that low expression of *MT-1H* induces cell growth arrest, retards cell migration, and suppresses prostate cancer cell invasion (Han et al. 2013). On the other hand, hypermethylation of *MT-1G* has been found to result in an increase in the aggressiveness of lesions and is characteristic of high-grade cancers (Henrique et al. 2005). Similarly, many other studies documented also the induction of MT expression by Zn exposure, including prostate cancer cell lines. MT-1/2 overexpression has been found after Zn induction, and correlated with an increase in the chemoresistance of the tumor as shown in neoplastically changed tissue as well as in cell lines (Smith et al. 2006; Krizkova et al. 2012; Gumulec et al. 2014). In prostate cancer cells, the mRNA levels of *MT-1A, MT-1X*, and *MT-2A* increased following Zn exposure (Hasumi et al. 2003). Interestingly, the expression of those isoforms previously described as nonfunctional—*MT-1J* and *MT-1M*—has been found to be inducible (Lin et al. 2009). Surprisingly, an effect analogous to metal exposure was observed under hypoxic conditions since the upregulation of *MT-1X* and *MT-2A* was found in human prostate cancer cell lines (LNCaP and PC-3) that were hypoxic, which is a typical state for neoplastic cells. Moreover, Yamasaki et al. found that the chemoresistance induced by this pathway can be diverted by siRNA, which may decrease *MT-2A* expression; this technique may possibly be useful as a treatment of cellular chemoresistance (Yamasaki et al. 2007).

4.11 Organs of the Female Reproductive Tract

4.11.1 Uterine Cervix

The expression of MT-1/2 has been demonstrated in preinvasive and invasive cervical squamous carcinoma (McCluggage et al. 1998). The pro-proliferative activity of MT-1/2 in cancer cells has been confirmed by strong Ki-67 expression colocalization. This points to the role of MT-1/2 overexpression in cervical intraepithelial neoplasia (CIN, preinvasive cancer) and in malignant lesions that may be related to cell proliferation (McCluggage et al. 1998). However, increased expression of MT-1/2 has been found in cervical cancer cells, with the characteristic scheme of staining observed in cancer nests. It was shown that hypoxic cells, mostly localized in the center of the tumor, did not express MT-1/2, or if they did, the reaction was very weak (Raleigh et al. 2000; Azuma et al. 2003). Moreover,

Raleigh et al. stated that inhibition of MT-1/2 expression in hypoxic cells was observed only in tissue studies, while in vitro study of cervical cancer cell lines showed increased MT-1/2 expression (Raleigh et al. 2000). Azuma et al. analyzed mRNA and protein levels, suggesting that the localization of MT-2A in the outer rims of the tumor nests was altered by differentiation and was strongly connected with transcriptional status of MT-1/2 (Azuma et al. 2003).

The phenomenon of tumor microenvironment remodeling is a known mechanism in the progression of neoplastic diseases. Many researchers have sought for reliable markers of this process. Based on a preliminary study (Dutsch-Wicherek et al. 2010) suggesting MT-1/2 expression associations with the abovementioned process, a precise study analyzing the microenvironment of cervical cancer has been performed (Walentowicz-Sadlecka et al. 2013). MT-1/2 expression was observed in cancer cells and in fibroblasts (CAFs, cancer-associated fibroblasts), as well as in macrophages (TAMs, tumor-associated macrophages) residing within the tumor microenvironment. The expression in cancer cells was increased through the whole nest, contrary to the results of Raleigh et al. and Azuma et al., but the hypoxic status was not checked. The expression of MT-1/2 in cancer cells, CAFs, and TAMs correlated positively with the stage of clinical advancement of FIGO (International Federation of Gynecology and Obstetrics) (Walentowicz-Sadlecka et al. 2013). These results seemed to be very informative and crucial for the proper choice of the treatment option—surgery. Moreover, they may possibly have some prognostic potential, since MT-1/2 expression correlated with local and distant advancement of the disease (Walentowicz-Sadlecka et al. 2013).

The primary means of treating cervical cancer is through surgical procedures; only in high clinical advancement it is supported by radiotherapy and chemotherapy. The chemoresistance profile in regard to MT-1/2 expression in cervical cancer is rather heterogeneous. While the occurrence of this cancer is strongly associated with HPV infection, heavy metal deposition, observed typically in the urothelium, has not been linked with cancerogenesis in this type of cancer. The analysis revealed increased intracellular free Zn concentrations in the Hep-2 cervical tumor cell line. It was postulated that Zn activates stress pathways that induce cellular demise. Increased levels of *MT-2A* expression induced by free Zn ions manifested in metal depositions in mitochondria and activated apoptosis, and in lysosomes led to necrosis (Rudolf and Cervinka 2010). A study on the immortalized human cervical cancer HeLa cells with chemically induced production of reactive oxygen species and apoptosis showed a significant increase in the mRNA expression of *MT-2A*, but not of *MT-1A* and *MT-2B*. These findings could confirm the inducible character of the MT-2A protein expression and its possible role in chemoresistance (Reinecke et al. 2006).

4.11.2 Endometrium

Immunohistochemical studies of MT-1/2 expression in the endometrium have demonstrated its presence in normal tissue, as well as in benign and malignant epithelial transformations—predominantly adenocarcinomas (McCluggage et al. 1999). Strong expression was found to be associated with high grade and advanced clinical stages of endometrial adenocarcinomas. Moreover, elevated MT-1/2 levels were typical of papillary serous-type adenocarcinomas (McCluggage et al. 1999). The MT-1/2 expression also correlated positively with the proliferation index and p53 expression and disclosed an inverse association with receptor status (progesterone and estrogen) (Ioachim et al. 2000). The findings suggested hormonal control of MT-1/2 expression in the normal endometrium, which could possibly be modified by p53 expression; its usage as an additional biological marker indicating aggressive behavior in endometrial lesions was also suggested (Ioachim et al. 2000). In regard to the various roles of MTs, including involvement in carcinogenesis, it was proposed that their function may be dependent on many additional factors, e.g., intracellular localization (Pedersen et al. 2009). Since nuclear localization of MT-1/2 had antiapoptotic and protective effects (Levadoux-Martin et al. 2001), it has been proposed that nuclear MT-1/2 overexpression in neoplastic cells may be associated with their increased survival and resistance to external factors, including radiotherapy and chemotherapy (Levadoux-Martin et al. 2001). The cytoplasmic expression of MT does not seem to differentiate between normal and neoplastic endometrium (when the secretory phase and adenocarcinoma cases where compared), whereas the nuclear MT expression was significantly stronger in cancer cells (Wicherek et al. 2005).

Aberrant gene expression of MT has been observed in several human tumors. Tse et al. studied the epigenetic alteration of *MT-1E* gene in human endometrial carcinomas. Microarray analysis showed a much lower expression of *MT-1E* in endometrial cancer cells than in other types of cancers, such as cervical, ovary, and prostate neoplasms. This result was confirmed by quantitative RT-PCR analysis of the *MT-1E*. The authors found a high frequency of cancer-specific hypermethylation of *MT-1E* and stated that the treatment of endometrial cancer cells with 5-azacytidine could potentially reactivate *MT-1E* expression (Tse et al. 2009).

4.11.3 Ovary

The expression of MT-1/2 differs between the normal and neoplastic ovarian epithelium. The overexpression of MT-1/2 in malignant ovarian surface epithelial tumors is considered to play a role in tumorigenesis. Analogous to endometrial carcinomas, there is a tendency toward its higher expression in poorly differentiated tumors (McCluggage et al. 2002). MT gene and protein expressions were associated

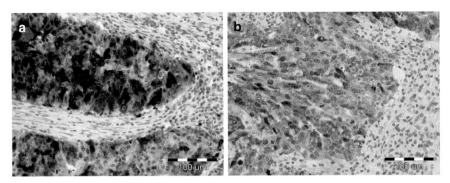

Fig. 4.6 *Immunohistochemical demonstration of metallothionein 1/2 in the cytoplasm and nuclei of endometrial* (**a**) *and ovarian adenocarcinoma* (**b**). Archival sections from the Department of Histology and Embryology, Wroclaw Medical University, Wroclaw, Poland

with the grade of histological malignancy since the increased expression of MT was detected in low-grade cancers with increased progression (Hengstler et al. 2001). A correlation between the size of the nuclei of cancer cells and the progression was found (especially in mucinous and serous ovarian tumors) (Fig. 4.6) (Tan et al. 1999). It was shown that MT could be a marker that characterized aggressiveness and high malignant potential in ovarian epithelial tumors (Zagorianakou et al. 2006). Moreover, it was suggested that in diagnostic problems, MT-1/2 may be helpful in differentiating between benign, borderline, and malignant tumors (Tan et al. 1999; Zagorianakou et al. 2006). Immunohistochemical tests revealed a set of antiapoptotic proteins, c-kit, telomerase, and MT-1/2, showing that p53 and MT-1/2 may be helpful in typing borderline and malignant ovarian tumors (Ozer et al. 2012). Additionally, the overexpression of survivin and MT-2A in recurrent ovarian tumors was analyzed (Tarapore et al. 2011). It was found that the knockdown of the two genes tested significantly decreased cellular proliferation in aggressive ovarian tumors and, therefore, it was suggested that targeting survivin and MT-2A genes by siRNA should be regarded as a potential therapeutic treatment (Tarapore et al. 2011). In view of the abovementioned studies it was hypothesized that ribozyme-induced downregulation of *MT-2A* expression resulted in activation of apoptosis in ovarian cancer cell lines (Tekur and Ho 2002). Additionally, studies on human ovarian surface epithelium (HOSE) with IL-1α-induced inflammation, which typically occurs during ovarian cancerogenesis, revealed that elevated levels of MT-2A expression are strongly associated with inflammation (Rae et al. 2004).

As with the observations of endometrial cancer, the analysis of separate intracellular fractions of MT-1/2 protein seems to be crucial since cytoplasmic MT-1/2 expression was found to be significantly higher in advanced ovarian cancer stages (III+IV vs. I+II) (Kobierzycki et al. 2013). However, preliminary studies did not demonstrate a relation between the expression of MT-1/2 and the intensity of HOSE proliferation before or after chemotherapy. The analysis of separate cellular MT-1/2 fractions demonstrated that exposure to cisplatin was paralleled by increasing

MT-1/2 expression in cell nuclei. The nuclear expression of MT-1/2 has also been found to be specific for ovarian cancers of poor clinical outcome (Surowiak et al. 2007). It was even shown that the lack of nuclear expression of MT-1/2 correlated with better progression-free survival (Woolston et al. 2010). No relationship was demonstrated between cytoplasmic MT-1/2 expression and clinical features, whereas nuclear expression was induced by cisplatin and seemed to protect DNA in the cells from the toxic effects of the drug (Surowiak et al. 2007). In view of the clinical outcome, the observations of Surowiak et al. confirm the results of the study by Hengstler et al. (2001). It was found that high MT-1/2 expression can be observed in low-grade carcinomas with an increased risk of progression (Hengstler et al. 2001; Surowiak et al. 2007). These results were confirmed by molecular studies showing that the overexpression of MT-2A is a typical feature of ovarian cancer cell lines that are resistant to cisplatin (Perego et al. 1998; Tarapore et al. 2011).

4.12 Central Nervous System

In the central nervous system, the expression of MT-1/2 has been found only in glial cells, while MT-3 localization has been predominantly observed in neurons (Masters et al. 1994; Nakajima and Suzuki 1995). At present, MT functions are mostly studied in neurodegenerative diseases and brain injuries, which are associated with MT-3, while oncological brain diseases are poorly described and paid particular attention to the relationships between MT-1/2 and chemoresistance aspects (Sogawa et al. 2001). MT-1/2 expression has been observed in regular as well as in neoplastically altered glial cells (Tews et al. 2000). The subcellular localization of MT-1/2 expression differs during the physiological activity of the glial cell. It has been shown in cell lines (HaCaT and C6) that the expression of MT-1/2 is initially localized in cytoplasm, and following 24 h induction by Cd ions, was subsequently observed in the nucleus (Nzengue et al. 2009). MT-1 and MT-2 can be found in both benign and malignant neoplasms, but their increased expression correlated with the increasing aggressive character of the lesions (Maier et al. 1997; Floriańczyk et al. 2003). Their expression is stronger in high-grade tumors than in low-grade ones, analogous to the expression of the proliferation antigen Ki-67. It has been postulated that MTs play a significant role in the growth of glial tumors, especially of astrocytic neoplasms (Hiura et al. 1998). Moreover, the decreased expression of MT-1/2 has been shown to be significantly associated with shorter progression-free survival time, while MT-1/2 increase was associated with shortened recurrence-free survival time (Korshunov et al. 1999). Interestingly, Tews et al. found that MT-1/2 is strongly expressed in blood vessels of normal and neoplastically changed tissues, which may possibly be responsible for the impaired penetration of chemotherapeutics into intact tissues (Tews et al. 2000).

As MT-1/2 expression has been found to be especially expressed in malignant brain tumors, it was anticipated that it would also be strongly associated with poor response to antineoplastic drugs (Maier et al. 1997). The astroglial cells that structurally and nutritionally support neurons represent the major sites for the accumulation and immobilization of toxic metal ions most probably connected with MTs due to the high cytosolic levels of MT 1, 2, and 3. The inducibility and metal binding of MTs were assessed in two human astrocytoma cell lines, U87 MG (grade IV) and IPDDC-2A (grade II), on exposure to cadmium chloride. It was found that Cd-dependent induction influenced MT1/2 expression only (Znidaric et al. 2007). Ryu et al. stated that MT-1E can modulate the motility and invasiveness of a human glioma cell line. This may involve two mechanisms: MT-1E may induce the changes in glioma cell morphology via the modification of cytoskeletal proteins, or it may modulate the activity of MMP-9 by NF-κB in these cell lines (Ryu et al. 2012). This is also in line with a report, which showed that MT-2A promotes breast cancer cell invasion by upregulating MMP-9 via AP-1 and NF-κB (Kim et al. 2011).

The expression of the MTs was also analyzed with respect to the chemotherapy of central nervous system malignancies, mostly gliomas. Primarily, chemoresistance against platinum-based chemotherapy compounds was found (Endresen et al. 1983; Doz et al. 1993). Subsequent studies have shown that overexpression of MTs can also confer resistance against many alkylating agents, demonstrating that specific thiol groups in MT-1/2 form covalent bonds with currently used chemotherapeutic drugs, such as chlorambucil, melphalan, cyclophosphamide, and mechlorethamine (Endresen et al. 1983; Kelley et al. 1988; Doz et al. 1993; Yu et al. 1995; Zaia et al. 1996; Wei et al. 1999). It was demonstrated that MT-1/2 may contribute to the resistance to the alkylating drug BCNU [1,3-bis (2-chloroethyl)-1-nitrosourea] by sequestering the drug's decomposition products (Bacolod et al. 2009). Falnoga et al. conducted a study to determine the influence of arsenic on the direct isoforms of MTs in malignant astrocytomas. In the astrocytoma cell line (U87 MG), it was shown that the increase in *MT-1X, MT-1F,* and *MT-2A* mRNA expression might be responsible for acquired resistance to arsenic-derivative chemotherapy (Falnoga et al. 2012). Further studies to determine the roles of the specific MT isoforms are necessary, especially in view of MT-associated chemoresistance.

4.13   Sarcomas

This broad group of malignancies of mesenchymal origin has not been widely analyzed in regard to MT expression. The few published studies have not shown consistent results. In sarcoma cells, the expression of MT-1/2 has been demonstrated in nuclei and in cytoplasm (Fig. 4.7) (Dziegiel et al. 2005). Analyzing soft tissue sarcomas (malignant fibrous histiocytoma, liposarcoma, and synovial sarcoma), a strong association between the expression of MT-1/2, Ki-67 antigen, and the grade of histological malignancy was observed (Dziegiel et al. 2005). Furthermore, patients with stronger MT-1/2 expression demonstrated shorter survival, suggesting the prognostic potential of MT-1/2 evaluation in sarcomas (Dziegiel

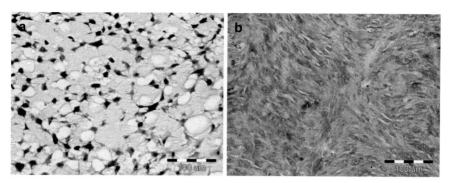

Fig. 4.7 *Immunohistochemical demonstration of metallothionein 1/2 in cytoplasm of liposarcoma* (**a**) *and fibrosarcoma* (**b**) *cells.* Archival sections from the Department of Histology and Embryology, Wroclaw Medical University, Wroclaw, Poland

et al. 2002; Dziegiel et al. 2005). MTs have also been studied in GIST and leiomyosarcoma (LMS). The overexpression of MT-1/2 in GIST and P-glycoprotein (P-gp) in LMS was shown to be associated with worse patient outcome (Perez-Gutierrez et al. 2007). These two proteins were also analyzed immunohistochemically by Shnyder et al. in osteosarcoma cell lines. Considering P-gp involvement in osteosarcoma chemoresistance, no association between P-gp and MT-1/2 on the protein level and no influence on clinical impact were found (Shnyder et al. 1998). Similar conclusions were reached by Endo-Munoz et al., who found upregulation of *MT-2A* and *MT-1E, MT-1X, MT-1H, MT-1B, MT-1G, MT-1L,* but no correlation between their expression levels and the response to chemotherapy (Endo-Munoz et al. 2010). At the molecular level in osteosarcoma cell lines, MT-2A overexpression by lentiviral transduction reduced bioavailable zinc levels, affected the reduction of osteosarcoma cell viability, and enhanced cell differentiation. On the other side, MT-2A silencing did not modify cell viability, but strongly inhibited the expression of osteoblastic markers and differentiation processes (Habel et al. 2013). The results suggest that MT-2A may be a potential prognostic marker for osteosarcoma sensitivity to chemotherapy partially mediated by zinc chelation (Habel et al. 2013). These divergent results clearly point to the need for further studies, mostly concerning the clinical outcome in regard to MT expression.

Chapter 5
Metallothioneins and Immune Function

5.1 Introduction

Early studies analyzing the expression of MTs in particular organs had already suggested the existence of a tissue-specific repertoire of MT proteins (Chakrabarty and Maiti 1985). MTs were also detected in immune organs, including the spleen, the bone marrow, and the thymus (Savino et al. 1984; Huber and Cousins 1993). In the human thymus, MTs were found to colocalize with thymulin on the cellular level (Savino et al. 1984). The biological activity of thymulin—a metallopeptide known to be one of the main regulators of immune cell development and function (Safieh-Garabedian et al. 2012)—strongly depends on zinc ion binding (Dardenne et al. 1982), which suggests a potential role of MTs as zinc donors in the control of active thymulin secretion. The expression of both the thymic and the bone marrow MT-1 and MT-2 genes has been shown to respond positively to inflammatory stimuli, such as IL-1, and to correlate with increased zinc uptake in these immune organs (Cousins and Leinart 1988). Multiple factors, including metal ions and proinflammatory substances, such as bacterial products (lipopolysaccharide, LPS) and mitogens (concanavalin A, Con A), have been demonstrated to stimulate MT expression in human peripheral blood immune cells, both directly and by inducing the secretion of other inflammatory mediators (Oberbarnscheidt et al. 1988; Vandeghinste et al. 2000). The basal and induced expression of MTs in immune cells has been shown to be dependent on the cell type, with monocytes expressing more MTs than lymphocytes and granulocytes (Harley et al. 1989; Yurkow and DeCoste 1999). In the lymphocytic population, cadmium-induced MT levels determined by a specific antibody and flow cytometry were shown to be higher in CD4+ and CD8+ T cells than in B cells and natural killer (NK) cells (Yurkow and Makhijani 1998). More detailed analyses performed on human peripheral blood lymphocytes have shown the expression of genes of various MT-isoforms, including *MT-1A*, *MT-1E*, *MT-1F*, *MT-1G*, *MT-1H*, *MT-1X*, *MT-2A*, and *MT-3*. The presence of MT-1A, MT-1E, MT-1F, MT-1G, MT-1H, MT-1X, MT-1K, and MT-2A has also been confirmed on the protein level (Vandeghinste et al. 2000).

Multiple components of the immune system, as well as cellular and humoral factors involved in various types of immune reaction, have been suggested as stimulators of MT expression. It has been demonstrated in chickens that IL-1, as an element of immune reaction to allogenic or bacterial stimulation, affects metal ion (Fe and Zn) metabolism via mechanisms such as overexpression of liver MTs and metal ion sequestration (Klasing 1984). In rodents, multiple cytokines, including IL-1, IL-6, and TNF-α, have been shown to induce MT expression in a cytokine-specific, MT-isoform, time- and organ-dependent pattern (Huber and Cousins 1993; Sato et al. 1994; Snyers and Content 1994).

5.2   Immune System Function

The influence of various essential and nonessential metals on the immune system has been extensively documented (Himeno et al. 2009; Fukada et al. 2011). The metal-binding properties of MTs have been investigated in studies on the role of metal cations in numerous physiological and pathological processes, including various forms of inflammation (Glennas 1983; Hartmann et al. 1985). Peripheral blood leukocytes are able to bind external metal ions, such as copper, to thioneins, with different efficiencies, i.e., monocytes are more effective than granulocytes (Hartmann et al. 1989). It has been demonstrated in human monocytic cell lines that metal cations (zinc, mercury, cadmium), even at nontoxic concentrations, induced MT protein accumulation, which was associated with the decreased ability of these cells to respond to proinflammatory stimulation with oxidative burst and IL-1b overexpression (Koropatnick and Zalups 1997). Apart from their intracellular localization, the presence of MT proteins has been described in numerous extracellular compartments, including various body fluids in physiological and pathological conditions (Milnerowicz et al. 2004; Swierzcek et al. 2004; Gumulec et al. 2012; Sliwinska-Mosson et al. 2012). Although MTs do not contain hydrophobic secretory signal sequences and, in many cases, extracellular MTs may result from cell leakage (Hidalgo et al. 1991), there are also data that suggest actual secretion of MT-1 from enterocytes, adipocytes, or astrocytes, probably through a nonclassical secretory system (Moltedo et al. 2000; Trayhurn et al. 2000; Klassen et al. 2004). Interestingly, immune cells have been demonstrated to release copper thioneins, which could in turn influence other immune and nonimmune cells (Hartmann et al. 1989).

Increasing evidence suggests that extracellular MTs may exert immune regulatory actions in ways other than as metal ion donors (Fig. 5.1). It has been shown that MT-1/2 can bind to the plasma membranes of murine macrophages, T cells, B cells, and cells of nonimmune origin, which implies the existence of a specific membrane receptor—possibly associated with the intracellular signal transduction system (Youn et al. 1995; Borghesi et al. 1996; El Refaey et al. 1997; Canpolat and

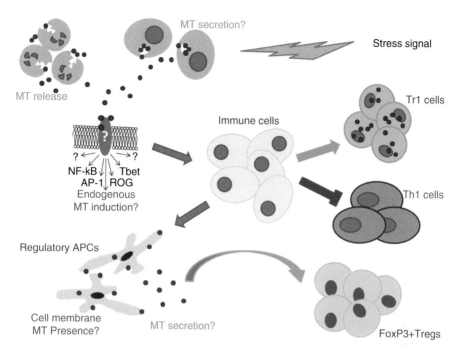

Fig. 5.1 *Effects of metallothioneins on immunoregulatory processes.* Stress signals may induce expression of both intracellular and extracellular MTs in the microenvironment of immune reaction. In consequence, regulatory cell populations of immune and nonimmune origin are induced and/or activated. Paracrine function of MT and interaction of soluble and/or cell membrane-bound MT molecule with its putative receptor represents one of the possible mechanisms of immunoregulatory activity of MTs. *AP-1* activator protein 1, *APCs* antigen-presenting cells, *FoxP3+Tregs* Fox3 (+) regulatory T cells, *NF-κB* nuclear factor kappa-light-chain-enhancer of activated B cells, *ROG* repressor of GATA, *Tbet* transcription factor of Th1, *Th1* type 1 helper T cells, *Tr1* type 1 regulatory T cells

Lynes 2001). The functional meaning of the putative MT-cell membrane receptor was supported by the finding that cells of immune origin (such as splenocytes, Jurkat T cells, and the macrophage-like cell line) elicited a chemotactic response to exogenous Zn/Cd-MT-1/2 applied in concentrations similar to those found in blood (Yin et al. 2005). Since cholera and pertussis toxin efficiently block the observed chemotactic activity, it can be assumed that the protein G pathway takes part in intracellular signaling activated by the MT receptors (Yin et al. 2005). So far, the best characterized candidate for an MT receptor is megalin, a member of the LDL receptor superfamily, also known as lipoprotein receptor-related protein or LRP2 (West et al. 2011). In their experimental study, Klaasen and colleagues provided several lines of evidence for the direct interaction of different MT-1 preparations with megalin, and of the crucial role of this process in MT-1 cellular internalization and renal heavy-metal uptake (Klassen et al. 2004). Furthermore, MT-1/2-associated neuroprotective activity has been shown to depend on MT-2A uptake by neurons mediated by megalin (and possibly by other LRP family receptors)

(Fitzgerald et al. 2007; Ambjorn et al. 2008; Chung et al. 2008a, b). Unfortunately, no data exist regarding the role of MT–LRP receptors in immune regulation.

It was demonstrated that, in both metal-bound forms and as apoproteins, MTs can significantly induce lymphocyte proliferation and regulate (positively or negatively) the proliferative response initiated by other proinflammatory substances such as LPS or Con A. Of great importance is the fact that the immune regulatory capacity of MTs depends on the type of applied proinflammatory stimuli and immune cell type, the quality of the complexed metal ions, and the measurable thiol level of the particular thionein (obtaining even an antiproliferative level in the case of Cu-MT and Hg-MT) (Lynes et al. 1990; Borghesi et al. 1996). Exogenous apo-MT-1/2 can regulate the differentiation of cytotoxic T lymphocytes in vitro, resulting in significantly lower activity against effector cells (Youn and Lynes 1999). In contrast, exogenous MT-1/2 enhances proliferative response and IL-2R expression in both allogeneic and autologous mixed leukocyte reaction accompanied by a significant decrease in CD8 and MHC I positive cells (Youn and Lynes 1999). These observations may be of particular interest in the context of cytotoxic immune reaction against tumor cells, which often express high MT levels (Pedersen et al. 2009; Krizkova et al. 2012). Copper thioneins could also inhibit the superoxide production of activated human blood phagocytes (Miesel et al. 1990). In contrast, Zn/Cd-MT-1/2 stimulates mouse peritoneal macrophages to produce increased levels of oxygen radicals, without altering their phagocytic activity. Accordingly, in another set of experiments, macrophages exposed to Zn/Cd-MT-1/2 showed a significantly increased ability to kill the engulfed yeast cells—an effect which could not be repeated by exposure to equimolar amounts of free zinc and cadmium (Youn et al. 1995). The regulatory effects of MTs on ROS production by immune cells were also observed in MT-1/2-null macrophages and in human monocytic cell lines with antisense-downregulated *MT-1* gene expression (Leibbrandt et al. 1994; Emeny et al. 2009). Furthermore, an abnormal costimulatory molecule expression and secretion profile of MT-deficient macrophages stimulated in vitro has been reported (Sugiura et al. 2004).

Extracellular MTs have also been demonstrated to possess regulatory properties in the humoral arm of the immune system. In an animal model, systemic administration of Zn/Cd-MT significantly reduced antigen-specific (ovalbumin, OVA) IgG response without affecting overall IgG production; this effect could be reversed by use of an anti-MT antibody, whereas free zinc and cadmium or apo-MT administration did not change the humoral immunity parameters in this system (Lynes et al. 1993). Similar effects of Zn/Cd-MT were observed in experiments investigating LPS-stimulated IgM production by splenocytes in vitro (Lynes et al. 1993). Concomitantly, in MT-1/2 knockout mice (MT-1/2-null), an active OVA immunization resulted in a significantly stronger specific humoral immune reaction, as evidenced by the increased in vivo anti-OVA IgG production, the higher number of Ig-secreting plasma cells in the spleen, and the appearance OVA-specific B cells (Crowthers et al. 2000). Significant differences were also found in the cellular composition of the spleen, thymus, and peripheral blood, with an apparent shift toward CD4+ and CD8+ T cells in the spleen of MT-1/2-null mice (Crowthers et al. 2000). Similar modulation of humoral response to OVA immunization has been obtained through systemic (intraperitoneal) administration of an MT-specific monoclonal antibody (mAb) to wild-type animals. Of importance seems to be the fact that the injection of the MT-mAb specifically increased the production of OVA-specific IgG1, but not of IgG2a, which implies modulation of Th2-dependent reaction (Canpolat and Lynes 2001). In contrast to the earlier study (Crowthers et al. 2000), other researches showed no gross abnormalities of the main immune cell populations in MT-1/2-null mice (Canpolat and Lynes 2001; Mita et al. 2002; Pankhurst et al. 2011).

Some not entirely consistent reports exist regarding the influence of MTs on the proliferative response of immune cells. The proliferative reaction of splenocytes from MT-1/2-null mice was shown to be impaired in response to mitogens (ConA and anti-CD3 Ab) but not to LPS stimulation. The observed decrease in proliferation was most probably specific to T cells and associated with lower IL-2 production by MT-1/2-null cells (Mita et al. 2002). Under normal conditions, exogenous MTs have been demonstrated to stimulate the proliferation of B cells, but not of T cells, as well as to increase the capacity of naive B cell differentiation into plasma cells (Borghesi et al. 1996). Since Zn/Cd-MT has also been shown to abrogate the proliferative effect exerted by macrophages on lymphocytes, it can be assumed that both endogenous and extracellular MTs can influence interactions between particular immune cell populations—modulating immune regulatory circuits in a manner dependent on inflammatory and stress conditions (Youn et al. 1995). In this aspect, extracellular MTs may act both as a regular part of stress response or—on the opposite—as a byproduct of cell damage. It has been speculated that extracellular MTs can induce the expression of the intracellular MT fraction and influence intracellular processes also in this way (McKim et al. 1992).

5.3 Immunoregulatory Mechanisms

In recent years, increasing attention has been paid to various immune and nonimmune cell populations that bear strong regulatory properties, as crucial elements of the physiology and pathology of the immune system (Stasiolek 2011; Lewinski et al. 2014). Among the most important immunoregulatory cell types are several regulatory T cell subpopulations (e.g., CD4+CD25+FoxP3+ T cells (Tregs) (Curotto de Lafaille and Lafaille 2009), IL-10-secreting type-1 regulatory T cells (Tr1) (Levings and Roncarolo 2000), TGF-β-secreting Th3 cells (Chen et al. 1994), regulatory CD8+ T cells (Sun et al. 1988), regulatory B cells (Dang et al. 2014), and various populations of dendritic cells (DCs) (including plasmacytoid and myeloid/conventional DCs) (Stasiolek et al. 2006; Stasiolek 2011; Kim and Diamond 2015), as well as anti-inflammatory monocytes and macrophages (Ziegler-Heitbrock 2007; Zhang et al. 2012).

The detailed analysis of specific regulatory immune cell populations in MT-1/2-null mice showed no changes in CD4+CD25+FoxP3+ Tregs (Huh et al. 2007;

Pankhurst et al. 2011). However, in MT-1/2-deficient mice, a significant impairment of T cell receptor-dependent naive T-cell differentiation into IL-10-secreting Tr1 cells was demonstrated. On the other hand, MT-1/2-null T cells differentiated more efficiently into IFN-γ-secreting cells under Th1 promoting stimulation (Huh et al. 2007). The role of MTs in Tr1 cell generation is further supported by experiments showing that exogenous MT-1/2 promotes the Tr1 secretion profile in wild-type CD4+ T cells upon TcR stimulation (Huh et al. 2007). In another set of experiments, it was demonstrated that peripheral blood mononuclear cells derived from MT-1/2-lacking animals are characterized by the significantly lower expression of the Ym1 molecule, which is a marker of the so-called alternatively activated macrophages—cells that differentiate under Th-2 type immune stimulation and exert a regulatory action important, for example, in antiparasite defenses (Goerdt and Orfanos 1999; Raes et al. 2002; Martinez et al. 2009; Pankhurst et al. 2011). Alternative macrophage polarization, as a mechanism of endotoxin tolerance induction, has been shown to be associated with profound changes in the expression of a broad panel of genes that code for immune active molecules (e.g., cytokines, chemokines, and growth factors). Most interestingly in this group, the expression of MT genes (*MT-1H*, *MT-1A*, *MT-1F*, *MT-2A*, *MT-1E*, *MT-1X*) showed very strong upregulation both in peripheral blood mononuclear cells and monocyte-derived macrophages (Pena et al. 2011).

MT-1 overexpression has also been demonstrated as an element of the genetic profile of tolerogenic mouse bone marrow-derived DCs induced by thermal stress and phytonutrient carvacrol treatment in vitro (Spiering et al. 2012). It has been recently shown that in human monocyte-derived DCs and mouse bone marrow-derived DCs, incubation with $ZnCl_2$ results in the upregulation of MT-1 protein expression and the presentation of MT-1 on the cell membrane as evidenced by flow cytometry (Spiering et al. 2014). Cell membrane MT-1-expressing mouse DCs were characterized by a regulatory phenotype and cytokine expression pattern and promoted FoxP3+ Tregs in coculture with T cells. Most interestingly, this effect was inhibited by anti-MT-1-blocking antibody. Functionally tolerogenic DCs could be also induced by glucocorticosteroid (dexamethasone) treatment; however, in this case, MT-1 expression increased only in the intracellular compartment and the tolerogenic DC properties were independent from anti-MT-1 antibody blocking (Spiering et al. 2014). It should be emphasized that this observation clearly indicates the existence of an active cell membrane receptor responsible for the MT-mediated interaction between immune cells (Fig. 5.1).

The immune dysfunction described in MT-1/2-null mice also encompasses abnormalities in the reaction of macrophages, microglia, and T cells to different kinds of brain injury (Penkowa et al. 1999a, b; Potter et al. 2007; Pankhurst et al. 2011). It is only in special cases that these observations may be explained by changes in the blood–brain barrier function in MT-1/2 animals. Most probably, the impairment of immunoregulatory MT function in these animals also affects local CNS inflammatory responses and the recruitment of cells from the periphery.

Genetic manipulation resulting in MT-1/2 deficiency has been shown to significantly augment autoimmune disorders in viable moth-eaten Hcph^{mev} (Lynes et al. 1999). Compared to wild-type animals, MT-1/2-null mice also showed greater susceptibility to an active form of experimental autoimmune encephalomyelitis (EAE) (Penkowa et al. 2001, 2003)—a disease mediated by myelin-specific T cells with lymphocytes secreting interferon IFN-γ (T helper type 1; Th1 cells) and IL-17 (Th17 cells), suggested to be crucial in the disease's induction (Gocke et al. 2007; O'Connor et al. 2008). Active EAE induced in MT-1/2-deficient animals was characterized by more pronounced CNS immune infiltration, increased accumulation of oxidative stress markers, and secretion of local proinflammatory cytokines (Penkowa et al. 2001; Penkowa and Hidalgo 2001). Unlike in the case of MT gene disruption, systemic (intraperitoneal) administration of Zn-MT-2 to EAE rats ameliorated the clinical symptoms, decreased macrophage and T lymphocyte recruitment, and decreased cytokine secretion in the affected CNS regions, while also reducing the number of macrophages in peripheral lymphatic organs (Penkowa and Hidalgo 2000). Significantly, some of the observations obtained in the EAE model performed with MT-1/2-null animals—the increased cytokine production, apoptosis rate, and ROS accumulation—were also observed in astrocytes (Penkowa et al. 2001), indicating that these regulatory "nonimmune" cells (Stasiolek 2011) are important elements of the immunoregulatory MT function in the CNS. Importantly, overexpression of MT genes was found by microarray analysis of acute and chronic multiple sclerosis (MS) lesions (Lock et al. 2002). Enhanced levels of MT-1 and MT-2 proteins have also been demonstrated in brain lesions of MS patients. The presence of MT proteins was found mainly in monocytes/macrophages and astrocytes, and was more pronounced in inactive lesions, suggesting the role of MTs in disease remission mechanisms (Penkowa et al. 2003).

In an animal model of collagen-induced arthritis (CIA)—another Th1-mediated autoimmune disease—MT-1-overexpressing transgenic animals exhibited weaker arthritis symptoms than did wild-type mice. Parallel to the clinical improvement, reduced serum levels of collagen-specific IgG and increased serum IL-10 levels were demonstrated in MT-transgenic animals, suggesting a positive influence of MTs on regulatory T cell generation in vivo (Huh et al. 2007).

A protective role for MTs against colonic mucosal inflammation has been suggested in various experimental colitis models. MT expression was shown to be upregulated in the acute phase of experimental colitis and downregulated during recovery (Devisscher et al. 2011). Intraperitoneal administration of MT-1 and MT-2 resulted in the amelioration of inflammatory damage to the colonic mucosa in experimental rat colitis (Altuner et al. 2010). Accordingly, acute colitis induced in MT-1/2-null mice was characterized by increased colonic mucosal production of proinflammatory cytokines (TNF-α, IFN-γ, and IL-17)—most probably dependent on infiltrating macrophages (Tsuji et al. 2013).

MT-1/2 immunofluorescence has also been demonstrated in thyroid follicular cells in Graves' disease, but not in Hashimoto's thyroiditis patients; however, there was no overt correlation between local thyroid MT expression and immune infiltration, antithyroid Ab titers, or clinical course of the autoimmune thyroid disease (Ruiz-Riol et al. 2012).

5.4 Apoptosis in the Immune System

The role of MTs in apoptotic processes represents another aspect of MT engagement in the regulation of immune system function. Programmed cell death (apoptosis), induced in target cells by some components of the immune system, is one of the key effector mechanisms of the inflammatory reaction. Strictly regulated apoptosis of immune effector cells in response to the constantly changing cellular and humoral microenvironment is also necessary for the correct outcome of the immune reaction (Lewinski et al. 2014). Indeed, even slight disturbances in the apoptosis program in immune cells may lead to protracted inflammatory responses and autoimmune diseases. Conversely, resistance to apoptosis in target cells is an immune evasion mechanism (Lewinski et al. 2014). Multiple lines of evidence suggest the close engagement of MTs in apoptosis regulation (Krizkova et al. 2012). The mechanism underlying the antiapoptotic activity of MT may involve direct interactions with transcription factors such as NF-κB or p53 tumor suppressor protein (Babula et al. 2012; Krizkova et al. 2012). Also, the influence on the p53 transcriptional activity or on the caspase cascade—through regulation of intracellular zinc content and/or oxidative stress processes—has been suggested as a possible mechanism of the apoptosis inhibiting action of MT (Chimienti et al. 2001; Raymond et al. 2010).

Various pathogens and pathogen-related products may modulate MT expression in different organs and cell types, including immune cells (Ilback et al. 2004; Zilliox et al. 2006; Frisk et al. 2007). Accordingly, in peripheral blood monocytes obtained from HIV patients during viremia, an increased expression of several MT-1 isoforms was observed (MT-1E, MT-1G, MT-1H, MT-1X) and correlated with the resistance of these cells to cadmium and FasL-induced apoptosis, through a mechanism most probably associated with increased intracellular zinc storage (Raymond et al. 2010). Splenocytes from MT-1/2-deficient animals were also significantly more prone to apoptosis in response to *Listeria monocytogenes* infection (Emeny et al. 2009). Experiments performed on MT-1/2-null mice showed a greater extent of programmed cell death of neurons and astrocytes in the course of autoimmune reaction to EAE (Penkowa et al. 2001). On the other hand, the systemic administration of Zn-MT-2 to EAE animals decreased the number of apoptotic neurons and oligodendrocytes and reduced the number of macrophages and lymphocytes (Penkowa and Hidalgo 2001). This may suggest the existence of different effector/target-cell-specific antiapoptotic mechanisms of extracellular MTs in the same immune reaction microenvironment. Besides the antioxidative action of MTs, the neuroprotective and antiapoptotic activity of exogenous MTs could possibly be dependent on the regulation of local cytokine secretion patterns and the overexpression of various growth factors—once again in a cell-specific manner with macrophages/microglia and astrocytes as the main sources of the Zn-MT-2-induced neurotrophic factors (Penkowa and Hidalgo 2001).

5.5   Endotoxin Shock

A convincing body of evidence associates MT expression with the mechanisms that underlie systemic reactions to bacterial toxins. Most probably, one of the major roles of MTs in immune pathology associated with endotoxin shock is the sequestration of metal ions (e.g., zinc). In mice lacking MT-1 and MT-2 expression, liver zinc concentration did not increase following LPS administration. Also, the typical rapid decrease in zinc plasma concentration was abrogated in LPS-treated MT-1/2-null mice, regardless of dosage (Philcox et al. 1995; Rofe et al. 1996). However, the pattern of MT isoform induction in response to endotoxin differed substantially from the changes stimulated by zinc or glucocorticosteroid administration. The effect of endotoxin on MT-1/2 expression in multiple organs (including the liver, kidneys, pancreas, intestine, lungs, heart, brain, ovary, uterus, and spleen) seems to be rather indirect and exerted by a complex network of secondary mediators induced in effector cells, including multiple cytokines, such as IL-1, IL-6, TNF-α, and IFN-γ (Abe et al. 1987; De et al. 1990). Out of these, IL-6 most probably plays a crucial role as an inductor of Stat1, Stat3, and MTF-1-dependent transcription of the *MT-1* gene (Lee et al. 1999; Hernandez et al. 2000). Similarly, in other animal models of toxic reactions to bacterial products, the overexpression of MT-1/2 in the liver in response to a bacterial exotoxin, toxic shock syndrome toxin-1 (TSST-1), was—to a great extent—secondary to the induction of the local secretion of cytokines, mainly IL-6, with no apparent engagement of IL-1 or TNF-α (Choudhuri et al. 1994). Interestingly, differences in MT expression in various mouse strains were suggested as molecular mechanism of susceptibility to systemic inflammatory response to LPS (De Maio et al. 1998). On the other hand, genetic MT overexpression induced in the target organs (liver, lungs, and kidneys) prior to LPS administration was associated with decreased peak plasma levels of TNF-a, IL-1b, IL-6, and TxB2 in the porcine model of endotoxemia (Klosterhalfen et al. 1996). Accordingly, in MT-1/2-null mice, the intraperitoneal administration of LPS was associated with increased organ and circulatory levels of multiple inflammatory factors (IL-1β, IL-6, GM-CSF, proinflammatory chemokines) and with increased inflammatory organ damage in lungs, kidneys, and liver (Inoue et al. 2006). MT-1/2-deficient mice were also shown to be more sensitive to LPS/GalN-induced lethality (Kimura et al. 2001). On the other hand, microarray analysis of the gene expression profile in pediatric septic shock has shown that genes of several MT isoforms (*MT-1E, MT-1M, MT-1G, MT-1H*) are overexpressed in peripheral blood obtained from nonsurvivors, as compared to the patients with a positive outcome. Additionally, serum zinc concentration in the nonsurvivor group was significantly lower, suggesting the influence of peripheral blood MTs on zinc systemic homeostasis in septic shock (Wong et al. 2007). In experimental settings, MT-1/2-null mice subjected to a murine polymicrobial sepsis model were characterized by better survival rate than wild-type animals (Wong et al. 2007). Moreover, MT-1/2-deficient mice also showed reduced sensitivity to TNF-α-induced toxicity in a model of lethal systemic inflammatory response syndrome, whereas mice overexpressing

MT-1 were more prone to the toxic effects of the intravenous TNF-α challenge (Waelput et al. 2001).

5.6 Oxidative Stress

As mentioned above, MTs are believed to exert complex antioxidative and thus cell-protective actions via a mechanism that involves both ROS scavenging and the inhibition of ROS production. Oxidative stress is believed to represent one of the main inducers of MT expression in the target organs (Andrews 2000). However, there are data suggesting that oxidants secreted in large amounts by neutrophils at the site of the inflammatory reaction may cleave the thiolate bonds of MTs and mobilize complexed zinc intracellularly, causing serious cellular harm (Fliss and Menard 1992). Nitric oxide (NO) is one of the most important mediators of the inflammatory process and may exert complex immunoregulatory effects (Stasiolek et al. 2000). One of the postulated mechanisms of NO activity is the depletion of effector cells at the site of the immune reaction. In a transgenic system, mouse fibroblast cell lines overexpressing the human *MT-1* gene showed significantly decreased sensitivity to NO-mediated cytotoxicity and nuclear DNA damage, probably through a mechanism involving the donation of thiols and the formation of iron–dinitrosyl complexes (Schwarz et al. 1995). The interaction between NO and MT thiol residues may, however, interfere with MT detoxification activity—for example, by releasing the MTs' bound cadmium, thus increasing the probability of cadmium-mediated DNA damage which may be of great significance in chronic inflammation and carcinogenesis (Misra et al. 1996).

5.7 Metal Toxicity

MTs have been shown to be crucially engaged in metal detoxification mechanisms. Since most known potentially toxic metal cations possess an affinity to MTs higher than that of zinc, it has been suggested that they displace zinc from the intracellular MT pool (Waalkes et al. 1984; Nielson et al. 1985; Sabolic et al. 2010). This process could profoundly alter the functional properties of MTs, and in turn contribute to the toxicity of particular metal cations. At the same time, the released zinc ions may increase MT expression through a mechanism associated with the direct activation and nuclear translocation of MTF-1 (Smirnova et al. 2000; McGee et al. 2010; Sabolic et al. 2010). However, non-zinc-associated mechanisms of MT induction by toxic metal cations have been also suggested, including the activation of inflammatory cytokine secretion (Daniels et al. 2002; Kobayashi et al. 2007). The overexpression of MTs in response to various stimulants has been found to be one of the protective mechanisms against cadmium cytotoxicity (Shiraishi et al. 1994; Sauer et al. 1997; Min et al. 2002), both in human peripheral blood monocytes and in T and B cells (Sone et al. 1988). This function of MTs seems to be dependent on the cells' activation state, since mitogen stimulation has been shown to alter the cadmium uptake of mouse splenic lymphocytes paralleled by MT IHC overexpression. Most importantly, mitogen stimulation also results in a marked redistribution of internalized cadmium from the nuclear to the cytoplasmic compartment, which is assumed to be an anticytotoxic mechanism (Sgagias et al. 1989). A similar cytosol-shift of the MT-bound cadmium has also been reported in other organs, as a protection mechanism against cadmium cytotoxicity (Goering and Klaassen 1983). Experiments performed on mice lacking the expression of MT-1/2 have confirmed the crucial role of MTs in mechanisms protecting against cadmium immunocytotoxicity. MT-1/2-null animals were distinctly more susceptible to cadmium-induced alteration in hematologic parameters, cytokine secretion, and changes in immune organs (spleen and thymus) morphology and function (Liu et al. 1999). Other potentially toxic metal cations (such as Pb and Hg) also affect MT expression in immune cells at their environmentally relevant concentrations. Interestingly, their toxic effects seem to be dependent on cell type and are more pronounced in the lymphocytic population than in monocytes (Fortier et al. 2008).

5.8   Role of MTs in the Resistance to Immune Therapy

As indicated above, MT expression is known to be induced by different forms of stress (Hidalgo et al. 1988a, b; Ghoshal et al. 1998; Hernandez et al. 2000). Glucocorticosteroids are among the main mediators of stress-induced MT expression (Hernandez et al. 2000), most probably via a mechanism involving the interaction of glucocorticosteroid receptors with the GRE in the promoter region of the MT genes (Kelly et al. 1997). It has been found that glucocorticosteroids influence the serum levels of MTs and induce MT expression in multiple organs and cell types—including liver, skin, amniotic cells, adipocyte, immune system, and cancer cell lines (Karin and Herschman 1980; Frings et al. 1989; Miesel and Zuber 1993; Snyers and Content 1994; Trayhurn et al. 2000; Dohi et al. 2005). On the contrary, under inflammatory conditions, glucocorticosteroids may counteract MT overexpression through a mechanism that most probably involves the suppression of proinflammatory cytokine secretion (Min et al. 1992). However, steroid therapy does not show any influence on the decreased expression of intestinal MT protein demonstrated in inflammatory bowel disease patients by means of IHC (Elmes et al. 1987). Additionally, the commonly used nonsteroidal anti-inflammatory drugs have been demonstrated to induce liver MT expression on the mRNA and protein level (Summer et al. 1989). These observations may be of great importance for patients with diseases that demand chronic anti-inflammatory therapy (Summer et al. 1989). Interestingly, the increased expression of MT-1 in macrophages and tubular epithelial cells is associated with glucocorticosteroid resistance in patients with acute renal allograft rejection (Rekers et al. 2013). The overexpression of MTs has been also suggested to be an important element of drug resistance mechanisms in metal ion-based and other anticancer drugs, as well as in chemical compounds used in primarily inflammatory conditions, such as rheumatoid arthritis (Glennas 1983; Dziegiel 2004). The induction of resistance to gold sodium thiomalate in a monocyte cell line has been associated with MT overexpression (Ichibangase et al. 1998). However, it was also suggested that increased presence of MTs may diminish the side effects of the applied therapy (Glennas 1983).

5.9   Intracellular Mechanisms

The exact molecular mechanism of the MT immune activity still needs further elucidation. Most interestingly, the increased DNA-binding activity of *NF-κB* and altered *NF-κB* subunit composition have been demonstrated in splenocytes from the OVA-immunized MT-null mice, as compared to the wild-type animals (Crowthers et al. 2000). The NF-κB transcription factor participates in the regulation of the expression of multiple molecules involved in virtually all known immune processes (such as cytokines, chemokines, immune active enzymes, adhesion molecules, and costimulatory molecules) (Ghosh et al. 1998). On the other hand, NF-κB transcriptional activity may be regulated in particular cell types by many different factors, such as metal ions, immune mediators, and oxidative stress products (Kim et al. 2003; Henkler et al. 2010; Hsieh and Yang 2013). The association of MTs and NF-κB activity seems to be complex and involves both indirect influences and direct interaction of particular MT isoforms with particular NF-κB subunits. The exact mechanisms of this interaction are not fully understood and seem to vary among different cell types and to depend on the cell compartment and cell functional state. Experiments performed on the MT-1/2-null embryonic cell lines, both transfected with *MT-1* gene and nontransfected, have shown that the TNF-α-induced DNA-binding activity of NF-κB is inhibited by MT-1 by means of a mechanism dependent on the decreased degradation of cytoplasmic NF-κB inhibitor (IκB) (Sakurai et al. 1999). A similar mechanism has been described in gastric cancer cell lines, where *MT-2A* overexpression was associated with increased expression of IκBα, decreased phosphorylation of the inhibitor, and thus the decreased nuclear localization of the p65 NF-κB subunit and the suppression of NF-κB transcriptional activity (Pan et al. 2013a). However, in another model, the transfection of human breast cancer cell line MCF-7 with the *MT-2A* gene resulted in the transactivation of NF-κB in the absence of additional stimulation (Takahashi et al. 2005). Although MTs do not represent typical nuclear localization signal (NLS)-mediated substrates (Nagano et al. 2000), their nuclear transport seems to be regulated by the oxidation of MTs' cytosolic protein partner in response to oxidative stress (Takahashi et al. 2005). Most interestingly, direct binding of the MT protein to the p50/RelA NF-κB complex (with p50 subunit as a direct target for MTs) has been documented in MCF-7 nuclear extracts upon stimulation with zinc ions (Abdel-Mageed and Agrawal 1998). Also, in the murine fibroblast-like cell line L929, the overexpression of MTs results in increased NF-κB binding activity, but the nuclear localization of the p50 NF-κB subunit in these experiments decreases (Kanekiyo et al. 2001). NF-κB activity may be also regulated on multiple levels by zinc ions (Haase and Rink 2014), which implies a possible engagement of MTs as modulators of cytoplasmic or nuclear zinc concentration (Kim et al. 1999, 2003). This possible mechanism is supported by the observation that the overexpression of the *MT-2A* gene in HeLa cells attenuates the increase in intracellular zinc levels upon zinc-ionophore treatment and concomitantly abrogates the inhibitory effect of zinc on NF-κB DNA-binding activity in these cells (Kim et al. 2003). The regulation of intracellular levels of oxidative stress products may represent yet another possible mechanism of MT interference with NF-κB and other transcription factor activity in activated immune cells (Rahman et al. 2004; Henkler et al. 2010; Hsieh and Yang 2013).

Other redox-sensitive intracellular signaling pathways may be also influenced by MTs. In experiments investigating the influence of MT-1/2 deficiency on naive T cell differentiation, it was shown that the apparent Th1 bias of MT-1/2-null T cells is associated with significantly increased binding activity of transcription factor activator protein 1 (AP-1) (Huh et al. 2007), a transcription factor engaged in multiple immune and growth processes, which in many cases interacts with the NF-κB pathway (Foletta et al. 1998; Gius et al. 1999). Moreover, under Th1-stimulating conditions, MT-1/2-null splenocytes expressed higher levels of Tbet—a transcription factor crucial for Th1 differentiation (Morinobu et al. 2004). On the other hand, upon TcR stimulation, the lower expression of repressor of GATA (ROG), the zinc finger transcriptional repressor engaged in the development of T helper populations, was observed (Miaw et al. 2000; Huh et al. 2007). In contrast, GATA3-transcription factor—which is involved in many T cell biological features, such as development, proliferation, and Th2 differentiation—is not affected by MT-1/2 deficiency (Huh et al. 2007; Wan 2014).

Acknowledgments

This work was funded by the scientific grant No. ST 540 of the Wroclaw Medical University. During manuscript preparation Bartosz Pula was supported by scholarship from the "Foundation for Polish Science."

References

Abdel-Mageed AB, Agrawal KC (1998) Activation of nuclear factor kappaB: potential role in metallothionein-mediated mitogenic response. Cancer Res 58(11):2335–2338

Abdel-Mageed AB, Zhao F, Rider BJ, Agrawal KC (2003) Erythropoietin-induced metallothionein gene expression: role in proliferation of K562 cells. Exp Biol Med (Maywood) 228(9):1033–1039

Abe S, Matsumi M, Tsukioki M, Mizukawa S, Takahashi T, Iijima Y, Itano Y, Kosaka F (1987) Metallothionein and zinc metabolism in endotoxin shock rats. Experientia Suppl 52:587–594

Abel J, de Ruiter N (1989) Inhibition of hydroxyl-radical-generated DNA degradation by metallothionein. Toxicol Lett 47(2):191–196

Ablett E, Whiteman DC, Boyle GM, Green AC, Parsons PG (2003) Induction of metallothionein in human skin by routine exposure to sunlight: evidence for a systemic response and enhanced induction at certain body sites. J Invest Dermatol 120(2):318–324

Adam V, Petrlova J, Wang J, Eckschlager T, Trnkova L, Kizek R (2010) Zeptomole electrochemical detection of metallothioneins. PLoS One 5(7), e11441

Adams L, Roth MJ, Abnet CC, Dawsey SP, Qiao YL, Wang GQ, Wei WQ, Lu N, Dawsey SM, Woodson K (2008) Promoter methylation in cytology specimens as an early detection marker for esophageal squamous dysplasia and early esophageal squamous cell carcinoma. Cancer Prev Res (Phila) 1(5):357–361

Ala S, Shokrzadeh M, Golpour M, Salehifar E, Alami M, Ahmadi A (2013) Zinc and copper levels in Iranian patients with psoriasis: a case control study. Biol Trace Elem Res 153(1–3):22–27

Albrecht AL, Singh RK, Somji S, Sens MA, Sens DA, Garrett SH (2008) Basal and metal-induced expression of metallothionein isoform 1 and 2 genes in the RWPE-1 human prostate epithelial cell line. J Appl Toxicol 28(3):283–293

Alkamal I, Ikromov O, Tölle A, Fuller TF, Magheli A, Miller K, Krause H, Kempkensteffen C (2015) An epigenetic screen unmasks metallothioneins as putative contributors to renal cell carcinogenesis. Urol Int 94(1):99–110

Alonso-Gonzalez C, Mediavilla D, Martinez-Campa C, Gonzalez A, Cos S, Sanchez-Barcelo EJ (2008) Melatonin modulates the cadmium-induced expression of MT-2 and MT-1 metallothioneins in three lines of human tumor cells (MCF-7, MDA-MB-231 and HeLa). Toxicol Lett 181(3):190–195

Altuner Y, Ayhanci A, Civi K, Ozden H, Ustuner D, Ustuner MC, Kurt H (2010) Protective effects of N(G)-nitro-L-arginine methyl ester and metallothioneins on excess nitric oxide toxicity in trinitrobenzene sulfonic acid-induced rat colitis. Anal Quant Cytol Histol 32(3):166–173

© Springer International Publishing Switzerland 2016
P. Dziegiel et al., *Metallothioneins in Normal and Cancer Cells*, Advances in Anatomy, Embryology and Cell Biology 218, DOI 10.1007/978-3-319-27472-0

Alves SM, Cardoso SV, de Fatima Bernardes V, Machado VC, Mesquita RA, Vieira do Carmo MA, Ferreira Aguiar MC (2007) Metallothionein immunostaining in adenoid cystic carcinomas of the salivary glands. Oral Oncol 43(3):252–256

Ambjorn M, Asmussen JW, Lindstam M, Gotfryd K, Jacobsen C, Kiselyov VV, Moestrup SK, Penkowa M, Bock E, Berezin V (2008) Metallothionein and a peptide modeled after metallothionein, EmtinB, induce neuronal differentiation and survival through binding to receptors of the low-density lipoprotein receptor family. J Neurochem 104(1):21–37

An J, Pan Y, Yan Z, Li W, Cui J, Yuan J, Tian L, Xing R, Lu Y (2013) MiR-23a in amplified 19p13.13 loci targets metallothionein 2A and promotes growth in gastric cancer cells. J Cell Biochem 114(9):2160–2169

Andrews GK (2000) Regulation of metallothionein gene expression by oxidative stress and metal ions. Biochem Pharmacol 59(1):95–104

Andrews PA, Murphy MP, Howell SB (1987) Metallothionein-mediated cisplatin resistance in human ovarian carcinoma cells. Cancer Chemother Pharmacol 19(2):149–154

Anezaki T, Ishiguro H, Hozumi I, Inuzuka T, Hiraiwa M, Kobayashi H, Yuguchi T, Wanaka A, Uda Y, Miyatake T et al (1995) Expression of growth inhibitory factor (GIF) in normal and injured rat brains. Neurochem Int 27(1):89–94

Apostolova MD, Choo KH, Michalska AE, Tohyama C (1997) Analysis of the possible protective role of metallothionein in streptozotocin-induced diabetes using metallothionein-null mice. J Trace Elem Med Biol 11(1):1–7

Apostolova MD, Ivanova IA, Cherian MG (1999) Metallothionein and apoptosis during differentiation of myoblasts to myotubes: protection against free radical toxicity. Toxicol Appl Pharmacol 159(3):175–184

Apostolova MD, Ivanova IA, Cherian MG (2000) Signal transduction pathways, and nuclear translocation of zinc and metallothionein during differentiation of myoblasts. Biochem Cell Biol 78(1):27–37

Apostolova MD, Chen S, Chakrabarti S, Cherian MG (2001) High-glucose-induced metallothionein expression in endothelial cells: an endothelin-mediated mechanism. Am J Physiol Cell Physiol 281(3):C899–907

Arai Y, Uchida Y, Takashima S (1997) Developmental immunohistochemistry of growth inhibitory factor in normal brains and brains of patients with Down syndrome. Pediatr Neurol 17 (2):134–138

Arriaga JM, Bravo IA, Bruno L, Morales Bayo S, Hannois A, Sanchez Loria F, Pairola F, Huertas E, Roberti MP, Rocca YS, Aris M, Barrio MM, Baffa Trasci S, Levy EM, Mordoh J, Bianchini M (2012a) Combined metallothioneins and p53 proteins expression as a prognostic marker in patients with Dukes stage B and C colorectal cancer. Hum Pathol 43 (10):1695–1703

Arriaga JM, Levy EM, Bravo AI, Bayo SM, Amat M, Aris M, Hannois A, Bruno L, Roberti MP, Loria FS, Pairola A, Huertas E, Mordoh J, Bianchini M (2012b) Metallothionein expression in colorectal cancer: relevance of different isoforms for tumor progression and patient survival. Hum Pathol 43(2):197–208

Arriaga JM, Greco A, Mordoh J, Bianchini M (2014) Metallothionein 1G and zinc sensitize human colorectal cancer cells to Chemotherapy. Mol Cancer Ther 13(5):1369–1381

Aschner M (1996a) Astrocytes as modulators of mercury-induced neurotoxicity. Neurotoxicology 17(3–4):663–669

Aschner M (1996b) The functional significance of brain metallothioneins. FASEB J 10 (10):1129–1136

Aschner M, Syversen T, Souza DO, Rocha JB (2006) Metallothioneins: mercury species-specific induction and their potential role in attenuating neurotoxicity. Exp Biol Med (Maywood) 231 (9):1468–1473

Athanassiadou P, Bantis A, Gonidi M, Athanassiades P, Agelonidou E, Grapsa D, Nikolopoulou P, Patsouris E (2007) The expression of metallothioneins on imprint smears of prostate carcinoma: correlation with clinicopathologic parameters and tumor proliferative capacity. Tumori 93(2):189–194

Azuma Y, Chou SC, Lininger RA, Murphy BJ, Varia MA, Raleigh JA (2003) Hypoxia and differentiation in squamous cell carcinomas of the uterine cervix: pimonidazole and involucrin. Clin Cancer Res 9(13):4944–4952

Babula P, Masarik M, Adam V, Eckschlager T, Stiborova M, Trnkova L, Skutkova H, Provaznik I, Hubalek J, Kizek R (2012) Mammalian metallothioneins: properties and functions. Metallomics 4(8):739–750

Back CM, Stohr S, Schafer EA, Biebermann H, Boekhoff I, Breit A, Gudermann T, Buch TR (2013) TSH induces metallothionein 1 in thyrocytes via Gq/11- and PKC-dependent signaling. J Mol Endocrinol 51(1):79–90

Bacolod MD, Fehdrau R, Johnson SP, Bullock NS, Bigner DD, Colvin M, Friedman HS (2009) BCNU-sequestration by metallothioneins may contribute to resistance in a medulloblastoma cell line. Cancer Chemother Pharmacol 63(4):753–758

Bahnson RR, Becich M, Ernstoff MS, Sandlow J, Cohen MB, Williams RD (1994) Absence of immunohistochemical metallothionein staining in bladder tumor specimens predicts response to neoadjuvant cisplatin, methotrexate and vinblastine chemotherapy. J Urol 152(6 Pt 2):2272–2275

Baird SK, Kurz T, Brunk UT (2006) Metallothionein protects against oxidative stress-induced lysosomal destabilization. Biochem J 394(Pt 1):275–283

Ballestín R, Blasco-Ibáñez JM, Crespo C, Nacher J, López-Hidalgo R, Gilabert-Juan J, Moltó D, Varea E (2014) Astrocytes of the murine model for Down Syndrome Ts65Dn display reduced intracellular ionic zinc. Neurochem Int 75:48–53

Banerjee D, Onosaka S, Cherian MG (1982) Immunohistochemical localization of metallothionein in cell nucleus and cytoplasm of rat liver and kidney. Toxicology 24(2):95–105

Barnes NL, Ackland ML, Cornish EJ (2000) Metallothionein isoform expression by breast cancer cells. Int J Biochem Cell Biol 32(8):895–903

Bedrnicek J, Vicha A, Jarosova M, Holzerova M, Cinatl J Jr, Michaelis M, Cinatl J, Eckschlager T (2005) Characterization of drug-resistant neuroblastoma cell lines by comparative genomic hybridization. Neoplasma 52(5):415–419

Bi Y, Palmiter RD, Wood KM, Ma Q (2004) Induction of metallothionein I by phenolic antioxidants requires metal-activated transcription factor 1 (MTF-1) and zinc. Biochem J 380 (Pt 3):695–703

Bieniek A, Pula B, Piotrowska A, Podhorska-Okolow M, Salwa A, Koziol M, Dziegiel P (2012) Expression of metallothionein I/II and Ki-67 antigen in various histological types of basal cell carcinoma. Folia Histochem Cytobiol 50(3):352–357

Bier B, Douglas-Jones A, Totsch M, Dockhorn-Dworniczak B, Bocker W, Jasani B, Schmid KW (1994) Immunohistochemical demonstration of metallothionein in normal human breast tissue and benign and malignant breast lesions. Breast Cancer Res Treat 30(3):213–221

Block TM, Mehta AS, Fimmel CJ, Jordan R (2003) Molecular viral oncology of hepatocellular carcinoma. Oncogene 22(33):5093–5107

Bonda DJ, Lee HG, Blair JA, Zhu X, Perry G, Smith MA (2011) Role of metal dyshomeostasis in Alzheimer's disease. Metallomics 3(3):267–270

Borges Junior PC, Ribeiro RI, Cardoso SV, Berbet AL, Rocha A, Espindola FS, Loyola AM (2007) Metallothionein immunolocalization in actinic skin nonmelanoma carcinomas. Appl Immunohistochem Mol Morphol 15(2):165–169

Borghesi LA, Youn J, Olson EA, Lynes MA (1996) Interactions of metallothionein with murine lymphocytes: plasma membrane binding and proliferation. Toxicology 108(1–2):129–140

Borthiry GR, Antholine WE, Kalyanaraman B, Myers JM, Myers CR (2007) Reduction of hexavalent chromium by human cytochrome b5: generation of hydroxyl radical and superoxide. Free Radic Biol Med 42(6):738–755, discussion 735-737

Borthiry GR, Antholine WE, Myers JM, Myers CR (2008) Addition of DNA to Cr(VI) and cytochrome b5 containing proteoliposomes leads to generation of DNA strand breaks and Cr (III) complexes. Chem Biodivers 5(8):1545–1557

Brandao R, Santos FW, Farina M, Zeni G, Bohrer D, Rocha JB, Nogueira CW (2006) Antioxidants and metallothionein levels in mercury-treated mice. Cell Biol Toxicol 22(6):429–438

Brown JJ, Xu H, William-Smith L, Mohamed H, Teklehaimanot S, Zhuo J, Osborne R, Liu F, Gowans RE, Nishitani J, Liu X (2003) Evaluation of metallothionein and p53 expression as potential prognostic markers for laryngeal squamous cell carcinoma. Cell Mol Biol (Noisy-le-grand) 49 Online Pub: OL473-479

Butcher HL, Kennette WA, Collins O, Zalups RK, Koropatnick J (2004) Metallothionein mediates the level and activity of nuclear factor kappa B in murine fibroblasts. J Pharmacol Exp Ther 310(2):589–598

Cai L, Satoh M, Tohyama C, Cherian MG (1999) Metallothionein in radiation exposure: its induction and protective role. Toxicology 132(2–3):85–98

Cai B, Zheng Q, Huang ZX (2005) The properties of the metal-thiolate clusters in recombinant mouse metallothionein-4. Protein J 24(6):327–336

Cai B, Zheng Q, Teng XC, Chen D, Wang Y, Wang KQ, Zhou GM, Xie Y, Zhang MJ, Sun HZ, Huang ZX (2006) The role of Thr5 in human neuron growth inhibitory factor. J Biol Inorg Chem 11(4):476–482

Campagne MV, Thibodeaux H, van Bruggen N, Cairns B, Lowe DG (2000) Increased binding activity at an antioxidant-responsive element in the metallothionein-1 promoter and rapid induction of metallothionein-1 and -2 in response to cerebral ischemia and reperfusion. J Neurosci 20(14):5200–5207

Cano-Gauci DF, Sarkar B (1996) Reversible zinc exchange between metallothionein and the estrogen receptor zinc finger. FEBS Lett 386(1):1–4

Canpolat E, Lynes MA (2001) In vivo manipulation of endogenous metallothionein with a monoclonal antibody enhances a T-dependent humoral immune response. Toxicol Sci 62 (1):61–70

Capdevila M, Domenech J, Pagani A, Tio L, Villarreal L, Atrian S (2005) Zn- and Cd-metallothionein recombinant species from the most diverse phyla may contain sulfide (S2-) ligands. Angew Chem Int Ed Engl 44(29):4618–4622

Cardoso SV, Barbosa HM, Candellori IM, Loyola AM, Aguiar MC (2002) Prognostic impact of metallothionein on oral squamous cell carcinoma. Virchows Arch 441(2):174–178

Cardoso SV, Silveira-Junior JB, De Carvalho MV, De-Paula AM, Loyola AM, De Aguiar MC (2009) Expression of metallothionein and p53 antigens are correlated in oral squamous cell carcinoma. Anticancer Res 29(4):1189–1193

Carrasco J, Giralt M, Molinero A, Penkowa M, Moos T, Hidalgo J (1999) Metallothionein (MT)-III: generation of polyclonal antibodies, comparison with MT-I+II in the freeze lesioned rat brain and in a bioassay with astrocytes, and analysis of Alzheimer's disease brains. J Neurotrauma 16(11):1115–1129

Cerulli N, Campanella L, Grossi R, Politi L, Scandurra R, Soda G, Gallo F, Damiani S, Alimonti A, Petrucci F, Caroli S (2006) Determination of Cd, Cu, Pb and Zn in neoplastic kidneys and in renal tissue of fetuses, newborns and corpses. J Trace Elem Med Biol 20 (3):171–179

Chakrabarty T, Maiti IB (1985) Immunological and biochemical properties of metallothioneins of golden hamster, mouse and rat. Mol Cell Biochem 68(1):41–47

Chan HM, Satoh M, Zalups RK, Cherian MG (1992) Exogenous metallothionein and renal toxicity of cadmium and mercury in rats. Toxicology 76(1):15–26

Chapman GA, Kay J, Kille P (1999) Structural and functional analysis of the rat metallothionein III genomic locus. Biochim Biophys Acta 1445(3):321–329

Chasapis CT, Loutsidou AC, Spiliopoulou CA, Stefanidou ME (2012) Zinc and human health: an update. Arch Toxicol 86(4):521–534

Chen Y, Kuchroo VK, Inobe J, Hafler DA, Weiner HL (1994) Regulatory T cell clones induced by oral tolerance: suppression of autoimmune encephalomyelitis. Science 265(5176):1237–1240

Chen H, Carlson EC, Pellet L, Moritz JT, Epstein PN (2001) Overexpression of metallothionein in pancreatic beta-cells reduces streptozotocin-induced DNA damage and diabetes. Diabetes 50 (9):2040–2046

Chen Y, Irie Y, Keung WM, Maret W (2002) S-nitrosothiols react preferentially with zinc thiolate clusters of metallothionein III through transnitrosation. Biochemistry 41(26):8360–8367

Chen WY, John JA, Lin CH, Lin HF, Wu SC, Chang CY (2004) Expression of metallothionein gene during embryonic and early larval development in zebrafish. Aquat Toxicol 69 (3):215–227

Cherian MG, Apostolova MD (2000) Nuclear localization of metallothionein during cell proliferation and differentiation. Cell Mol Biol (Noisy-le-grand) 46(2):347–356

Cherian MG, Kang YJ (2006) Metallothionein and liver cell regeneration. Exp Biol Med (Maywood) 231(2):138–144

Cherian MG, Jayasurya A, Bay BH (2003) Metallothioneins in human tumors and potential roles in carcinogenesis. Mutat Res 533(1–2):201–209

Chiaverini N, De Ley M (2010) Protective effect of metallothionein on oxidative stress-induced DNA damage. Free Radic Res 44(6):605–613

Chimienti F, Jourdan E, Favier A, Seve M (2001) Zinc resistance impairs sensitivity to oxidative stress in HeLa cells: protection through metallothioneins expression. Free Radic Biol Med 31 (10):1179–1190

Chin JL, Banerjee D, Kadhim SA, Kontozoglou TE, Chauvin PJ, Cherian MG (1993) Metallothionein in testicular germ cell tumors and drug resistance. Clinical correlation. Cancer 72(10):3029–3035

Choi CH, Cha YJ, An CS, Kim KJ, Kim KC, Moon SP, Lee ZH, Min YD (2004) Molecular mechanisms of heptaplatin effective against cisplatin-resistant cancer cell lines: less involvement of metallothionein. Cancer Cell Int 4(1):6

Choudhuri S, McKim JM Jr, Klaassen CD (1994) Induction of metallothionein by superantigenic bacterial exotoxin: probable involvement of the immune system. Biochim Biophys Acta 1225 (2):171–179

Choudhuri S, Kramer KK, Berman NE, Dalton TP, Andrews GK, Klaassen CD (1995) Constitutive expression of metallothionein genes in mouse brain. Toxicol Appl Pharmacol 131 (1):144–154

Chubatsu LS, Gennari M, Meneghini R (1992) Glutathione is the antioxidant responsible for resistance to oxidative stress in V79 Chinese hamster fibroblasts rendered resistant to cadmium. Chem Biol Interact 82(1):99–110

Chun JH, Kim HK, Kim E, Kim IH, Kim JH, Chang HJ, Choi IJ, Lim HS, Kim IJ, Kang HC, Park JH, Bae JM, Park JG (2004) Increased expression of metallothionein is associated with irinotecan resistance in gastric cancer. Cancer Res 64(14):4703–4706

Chung RS, Vickers JC, Chuah MI, Eckhardt BL, West AK (2002) Metallothionein-III inhibits initial neurite formation in developing neurons as well as postinjury, regenerative neurite sprouting. Exp Neurol 178(1):1–12

Chung RS, Hidalgo J, West AK (2008a) New insight into the molecular pathways of metallothionein-mediated neuroprotection and regeneration. J Neurochem 104(1):14–20

Chung RS, Penkowa M, Dittmann J, King CE, Bartlett C, Asmussen JW, Hidalgo J, Carrasco J, Leung YK, Walker AK, Fung SJ, Dunlop SA, Fitzgerald M, Beazley LD, Chuah MI, Vickers JC, West AK (2008b) Redefining the role of metallothionein within the injured brain: extracellular metallothioneins play an important role in the astrocyte-neuron response to injury. J Biol Chem 283(22):15349–15358

Colangelo V, Schurr J, Ball MJ, Pelaez RP, Bazan NG, Lukiw WJ (2002) Gene expression profiling of 12633 genes in Alzheimer hippocampal CA1: transcription and neurotrophic factor down-regulation and up-regulation of apoptotic and pro-inflammatory signaling. J Neurosci Res 70(3):462–473

Cong W, Ma W, Zhao T, Zhu Z, Wang Y, Tan Y, Li X, Jin L, Cai L (2013) Metallothionein prevents diabetes-induced cardiac pathological changes, likely via the inhibition of succinyl-CoA:3-ketoacid coenzyme A transferase-1 nitration at Trp(374). Am J Physiol Endocrinol Metab 304(8):E826–835

Conrad CC, Grabowski DT, Walter CA, Sabia M, Richardson A (2000) Using MT(-/-) mice to study metallothionein and oxidative stress. Free Radic Biol Med 28(3):447–462

Cousins RJ, Leinart AS (1988) Tissue-specific regulation of zinc metabolism and metallothionein genes by interleukin 1. FASEB J 2(13):2884–2890

Coyle P, Mathew G, Game PA, Myers JC, Philcox JC, Rofe AM, Jamieson GG (2002a) Metallothionein in human oesophagus, Barrett's epithelium and adenocarcinoma. Br J Cancer 87(5):533–536

Coyle P, Philcox JC, Carey LC, Rofe AM (2002b) Metallothionein: the multipurpose protein. Cell Mol Life Sci 59(4):627–647

Crowthers KC, Kline V, Giardina C, Lynes MA (2000) Augmented humoral immune function in metallothionein-null mice. Toxicol Appl Pharmacol 166(3):161–172

Cui Y, Wang J, Zhang X, Lang R, Bi M, Guo L, Lu SH (2003) ECRG2, a novel candidate of tumor suppressor gene in the esophageal carcinoma, interacts directly with metallothionein 2A and links to apoptosis. Biochem Biophys Res Commun 302(4):904–915

Curotto de Lafaille MA, Lafaille JJ (2009) Natural and adaptive foxp3+ regulatory T cells: more of the same or a division of labor? Immunity 30(5):626–635

Dalgin GS, Drever M, Williams T, King T, DeLisi C, Liou LS (2008) Identification of novel epigenetic markers for clear cell renal cell carcinoma. J Urol 180(3):1126–1130

D'Amico E, Factor-Litvak P, Santella RM, Mitsumoto H (2013) Clinical perspective on oxidative stress in sporadic amyotrophic lateral sclerosis. Free Radic Biol Med 65:509–527

Dang VD, Hilgenberg E, Ries S, Shen P, Fillatreau S (2014) From the regulatory functions of B cells to the identification of cytokine-producing plasma cell subsets. Curr Opin Immunol 28C:77–83

Daniels PJ, Bittel D, Smirnova IV, Winge DR, Andrews GK (2002) Mammalian metal response element-binding transcription factor-1 functions as a zinc sensor in yeast, but not as a sensor of cadmium or oxidative stress. Nucleic Acids Res 30(14):3130–3140

Dardenne M, Pleau JM, Nabarra B, Lefrancier P, Derrien M, Choay J, Bach JF (1982) Contribution of zinc and other metals to the biological activity of the serum thymic factor. Proc Natl Acad Sci USA 79(17):5370–5373

Datta J, Majumder S, Kutay H, Motiwala T, Frankel W, Costa R, Cha HC, MacDougald OA, Jacob ST, Ghoshal K (2007) Metallothionein expression is suppressed in primary human hepatocellular carcinomas and is mediated through inactivation of CCAAT/enhancer binding protein alpha by phosphatidylinositol 3-kinase signaling cascade. Cancer Res 67(6):2736–2746

Davis SR, Cousins RJ (2000) Metallothionein expression in animals: a physiological perspective on function. J Nutr 130(5):1085–1088

Davis SR, Samuelson DA, Cousins RJ (2001) Metallothionein expression protects against carbon tetrachloride-induced hepatotoxicity, but overexpression and dietary zinc supplementation provide no further protection in metallothionein transgenic and knockout mice. J Nutr 131 (2):215–222

De Maio A, Mooney ML, Matesic LE, Paidas CN, Reeves RH (1998) Genetic component in the inflammatory response induced by bacterial lipopolysaccharide. Shock 10(5):319–323

De SK, McMaster MT, Andrews GK (1990) Endotoxin induction of murine metallothionein gene expression. J Biol Chem 265(25):15267–15274

Deng D, El-Rifai W, Ji J, Zhu B, Trampont P, Li J, Smith MF, Powel SM (2003a) Hypermethylation of metallothionein-3 CpG island in gastric carcinoma. Carcinogenesis 24 (1):25–29

Deng D, El-Rifai W, Ji J, Zhu B, Trampont P, Li J, Smith MF, Powel SM (2003b) Hypermethylation of metallothionein-3 CpG island in gastric carcinoma. Carcinogenesis 24 (1):25–29

Devisscher L, Hindryckx P, Olievier K, Peeters H, De Vos M, Laukens D (2011) Inverse correlation between metallothioneins and hypoxia-inducible factor 1 alpha in colonocytes and experimental colitis. Biochem Biophys Res Commun 416(3–4):307–312

Ding ZC, Teng XC, Zheng Q, Ni FY, Cai B, Wang Y, Zhou GM, Sun HZ, Tan XS, Huang ZX (2009) Important roles of the conserved linker-KKS in human neuronal growth inhibitory factor. Biometals 22(5):817–826

Ding ZC, Ni FY, Huang ZX (2010) Neuronal growth-inhibitory factor (metallothionein-3): structure-function relationships. FEBS J 277(14):2912–2920

Dohi Y, Shimaoka H, Ikeuchi M, Ohgushi H, Yonemasu K, Minami T (2005) Role of metallothionein isoforms in bone formation processes in rat marrow mesenchymal stem cells in culture. Biol Trace Elem Res 104(1):57–70

Douglas-Jones AG, Schmid KW, Bier B, Horgan K, Lyons K, Dallimore ND, Moneypenny IJ, Jasani B (1995) Metallothionein expression in duct carcinoma in situ of the breast. Hum Pathol 26(2):217–222

Doz F, Roosen N, Rosenblum ML (1993) Metallothionein and anticancer agents: the role of metallothionein in cancer chemotherapy. J Neurooncol 17(2):123–129

Duncan KE, Ngu TT, Chan J, Salgado MT, Merrifield ME, Stillman MJ (2006) Peptide folding, metal-binding mechanisms, and binding site structures in metallothioneins. Exp Biol Med (Maywood) 231(9):1488–1499

Dutsch-Wicherek M, Popiela TJ, Klimek M, Rudnicka-Sosin L, Wicherek L, Oudinet JP, Skladzien J, Tomaszewska R (2005) Metallothionein stroma reaction in tumor adjacent healthy tissue in head and neck squamous cell carcinoma and breast adenocarcinoma. Neuro Endocrinol Lett 26(5):567–574

Dutsch-Wicherek M, Tomaszewska R, Lazar A, Strek P, Wicherek Ł, Piekutowski K, Jozwicki W (2010) The evaluation of metallothionein expression in nasal polyps with respect to immune cell presence and activity. BMC Immunol 11:10

Dutta R, Sens DA, Somji S, Sens MA, Garrett SH (2002) Metallothionein isoform 3 expression inhibits cell growth and increases drug resistance of PC-3 prostate cancer cells. Prostate 52 (2):89–97

Duval D, Trouillas M, Thibault C, Dembelé D, Diemunsch F, Reinhardt B, Mertz AL, Dierich A, Boeuf H (2006) Apoptosis and differentiation commitment: novel insights revealed by gene profiling studies in mouse embryonic stem cells. Cell Death Differ 13(4):564–575

Dziegiel P (2004) Expression of metallothioneins in tumor cells. Pol J Pathol 55(1):3–12

Dziegiel P, Suder E, Surowiak P, Kornafel J, Zabel M (2002) Expression of metallothionein in synovial sarcoma cells. Appl Immunohistochem Mol Morphol 10(4):357–362

Dziegiel P, Forgacz J, Suder E, Surowiak P, Kornafel J, Zabel M (2003) Prognostic significance of metallothionein expression in correlation with Ki-67 expression in adenocarcinomas of large intestine. Histol Histopathol 18(2):401–407

Dziegiel P, Dumanska M, Forgacz J, Wojna A, Zabel M (2004a) Intensity of apoptosis as related to the expression of metallothionein (MT), caspase-3 (cas-3) and Ki-67 antigen and the survival time of patients with primary colorectal adenocarcinomas. Rocz Akad Med Bialymst 49(Suppl 1):5–7

Dziegiel P, Jelen M, Muszczynska B, Maciejczyk A, Szulc A, Podhorska-Okolow M, Cegielski M, Zabel M (2004b) Role of metallothionein expression in non-small cell lung carcinomas. Rocz Akad Med Bialymst 49(Suppl 1):43–45

Dziegiel P, Salwa-Zurawska W, Zurawski J, Wojnar A, Zabel M (2005) Prognostic significance of augmented metallothionein (MT) expression correlated with Ki-67 antigen expression in selected soft tissue sarcomas. Histol Histopathol 20(1):83–89

Ebert MP, Gunther T, Hoffmann J, Yu J, Miehlke S, Schulz HU, Roessner A, Korc M, Malfertheiner P (2000) Expression of metallothionein II in intestinal metaplasia, dysplasia, and gastric cancer. Cancer Res 60(7):1995–2001

Eibl JK, Abdallah Z, Ross GM (2010) Zinc-metallothionein: a potential mediator of antioxidant defence mechanisms in response to dopamine-induced stress. Can J Physiol Pharmacol 88 (3):305–312

Eid H, Géczi L, Mágori A, Bodrogi I, Institoris E, Bak M (1998) Drug resistance and sensitivity of germ cell testicular tumors: evaluation of clinical relevance of MDR1/Pgp, p53, and metallothionein (MT) proteins. Anticancer Res 18(4C):3059–3064

El Refaey H, Ebadi M, Kuszynski CA, Sweeney J, Hamada FM, Hamed A (1997) Identification of metallothionein receptors in human astrocytes. Neurosci Lett 231(3):131–134

El Sharkawy SL, Farrag AR (2008) Mean nuclear area and metallothionein expression in ductal breast tumors: correlation with estrogen receptor status. Appl Immunohistochem Mol Morphol 16(2):108–112

El Sharkawy SL, Abbas NF, Badawi MA, El Shaer MA (2006) Metallothionein isoform II expression in hyperplastic, dysplastic and neoplastic prostatic lesions. J Clin Pathol 59 (11):1171–1174

Elmes ME, Clarkson JP, Jasani B (1987) Histological demonstration of immunoreactive metallothionein in rat and human tissues. Experientia Suppl 52:533–537

Emeny RT, Marusov G, Lawrence DA, Pederson-Lane J, Yin X, Lynes MA (2009) Manipulations of metallothionein gene dose accelerate the response to Listeria monocytogenes. Chem Biol Interact 181(2):243–253

Emri E, Egervari K, Varvolgyi T, Rozsa D, Miko E, Dezso B, Veres I, Mehes G, Emri G, Remenyik E (2013) Correlation among metallothionein expression, intratumoural macrophage infiltration and the risk of metastasis in human cutaneous malignant melanoma. J Eur Acad Dermatol Venereol 27(3):e320–327

Endo T, Yoshikawa M, Ebara M, Kato K, Sunaga M, Fukuda H, Hayasaka A, Kondo F, Sugiura N, Saisho H (2004) Immunohistochemical metallothionein expression in hepatocellular carcinoma: relation to tumor progression and chemoresistance to platinum agents. J Gastroenterol 39(12):1196–1201

Endo-Munoz L, Cumming A, Sommerville S, Dickinson I, Saunders NA (2010) Osteosarcoma is characterised by reduced expression of markers of osteoclastogenesis and antigen presentation compared with normal bone. Br J Cancer 103(1):73–81

Endresen L, Bakka A, Rugstad HE (1983) Increased resistance to chlorambucil in cultured cells with a high concentration of cytoplasmic metallothionein. Cancer Res 43(6):2918–2926

Erickson JC, Hollopeter G, Thomas SA, Froelick GJ, Palmiter RD (1997) Disruption of the metallothionein-III gene in mice: analysis of brain zinc, behavior, and neuron vulnerability to metals, aging, and seizures. J Neurosci 17(4):1271–1281

Faller P (2010) Neuronal growth-inhibitory factor (metallothionein-3): reactivity and structure of metal-thiolate clusters. FEBS J 277(14):2921–2930

Faller WJ, Rafferty M, Hegarty S, Gremel G, Ryan D, Fraga MF, Esteller M, Dervan PA, Gallagher WM (2010) Metallothionein 1E is methylated in malignant melanoma and increases sensitivity to cisplatin-induced apoptosis. Melanoma Res 20(5):392–400

Falnoga I, Zelenik Pevec A, Šlejkovec Z, Žnidarič MT, Zajc I, Mlakar SJ, Marc J (2012) Arsenic trioxide (ATO) influences the gene expression of metallothioneins in human glioblastoma cells. Biol Trace Elem Res 149(3):331–339

Fan LZ, Cherian MG (2002) Potential role of p53 on metallothionein induction in human epithelial breast cancer cells. Br J Cancer 87(9):1019–1026

Ferrario C, Lavagni P, Gariboldi M, Miranda C, Losa M, Cleris L, Formelli F, Pilotti S, Pierotti MA, Greco A (2008) Metallothionein 1G acts as an oncosupressor in papillary thyroid carcinoma. Lab Invest 88(5):474–481

Fic M, Pula B, Rogala K, Dziegiel P (2013) Role of metallothionein expression alimentary tract cancers. Post Biol Kom 40(1):5–20

Finn SP, Smyth P, Cahill S, Streck C, O'Regan EM, Flavin R, Sherlock J, Howells D, Henfrey R, Cullen M, Toner M, Timon C, O'Leary JJ, Sheils OM (2007) Expression microarray analysis of papillary thyroid carcinoma and benign thyroid tissue: emphasis on the follicular variant and potential markers of malignancy. Virchows Arch 450(3):249–260

Fitzgerald M, Nairn P, Bartlett CA, Chung RS, West AK, Beazley LD (2007) Metallothionein-IIA promotes neurite growth via the megalin receptor. Exp Brain Res 183(2):171–180

Fliss H, Menard M (1992) Oxidant-induced mobilization of zinc from metallothionein. Arch Biochem Biophys 293(1):195–199

Floriańczyk B, Kaczmarczyk R, Osuchowski J, Stryjecka-Zimmer M, Trojanowski T, Marzec Z (2003) Metallothioneins and microelements in brain tumours. Ann Univ Mariae Curie Sklodowska Med 58(1):1–4

Foletta VC, Segal DH, Cohen DR (1998) Transcriptional regulation in the immune system: all roads lead to AP-1. J Leukoc Biol 63(2):139–152

Fong LY, Jiang Y, Farber JL (2006) Zinc deficiency potentiates induction and progression of lingual and esophageal tumors in p53-deficient mice. Carcinogenesis 27(7):1489–1496

Formigari A, Irato P, Santon A (2007) Zinc, antioxidant systems and metallothionein in metal mediated-apoptosis: biochemical and cytochemical aspects. Comp Biochem Physiol C Toxicol Pharmacol 146(4):443–459

Formigari A, Gregianin E, Irato P (2013) The effect of zinc and the role of p53 in copper-induced cellular stress responses. J Appl Toxicol 33(7):527–536

Fortier M, Omara F, Bernier J, Brousseau P, Fournier M (2008) Effects of physiological concentrations of heavy metals both individually and in mixtures on the viability and function of peripheral blood human leukocytes in vitro. J Toxicol Environ Health A 71(19):1327–1337

Foster AW, Robinson NJ (2011) Promiscuity and preferences of metallothioneins: the cell rules. BMC Biol 9:25

Frederickson CJ, Moncrieff DW (1994) Zinc-containing neurons. Biol Signals 3(3):127–139

Friedline JA, Garrett SH, Somji S, Todd JH, Sens DA (1998) Differential expression of the MT-1E gene in estrogen-receptor-positive and -negative human breast cancer cell lines. Am J Pathol 152(1):23–27

Frings MM, Kind PP, Goerz G, Abel J (1989) The effect of topically applied corticosteroids and zinc on the metallothionein content of skin in an experimental model. Clin Exp Dermatol 14(6):434–436

Frisk P, Tallkvist J, Gadhasson IL, Blomberg J, Friman G, Ilback NG (2007) Coxsackievirus B3 infection affects metal-binding/transporting proteins and trace elements in the pancreas in mice. Pancreas 35(3):e37–44

Fu J, Lv H, Guan H, Ma X, Ji M, He N, Shi B, Hou P (2013) Metallothionein 1G functions as a tumor suppressor in thyroid cancer through modulating the PI3K/Akt signaling pathway. BMC Cancer 13(1):462

Fukada T, Yamasaki S, Nishida K, Murakami M, Hirano T (2011) Zinc homeostasis and signaling in health and diseases: Zinc signaling. J Biol Inorg Chem 16(7):1123–1134

Galizia G, Ferraraccio F, Lieto E, Orditura M, Castellano P, Imperatore V, La Manna G, Pinto M, Ciardiello F, La Mura A, De Vita F (2006) p27 downregulation and metallothionein overexpression in gastric cancer patients are associated with a poor survival rate. J Surg Oncol 93(3):241–252

Gallicchio L, Flaws JA, Sexton M, Ioffe OB (2004) Cigarette smoking and metallothionein expression in invasive breast carcinomas. Toxicol Lett 152(3):245–253

Gao Y, Han Z, Liu X (1997) Metallothionein expression and its significance in salivary gland tumors. Zhonghua Kou Qiang Yi Xue Za Zhi 32(5):282–284

Garrett SH, Sens MA, Shukla D, Nestor S, Somji S, Todd JH, Sens DA (1999) Metallothionein isoform 3 expression in the human prostate and cancer-derived cell lines. Prostate 41(3):196–202

Garrett SH, Sens MA, Shukla D, Flores L, Somji S, Todd JH, Sens DA (2000) Metallothionein isoform 1 and 2 gene expression in the human prostate: downregulation of MT-1X in advanced prostate cancer. Prostate 43(2):125–135

Garrett SH, Phillips V, Somji S, Sens MA, Dutta R, Park S, Kim D, Sens DA (2002) Transient induction of metallothionein isoform 3 (MT-3), c-fos, c-jun and c-myc in human proximal tubule cells exposed to cadmium. Toxicol Lett 126(1):69–80

Gauthier MA, Eibl JK, Crispo JA, Ross GM (2008) Covalent arylation of metallothionein by oxidized dopamine products: a possible mechanism for zinc-mediated enhancement of dopaminergic neuron survival. Neurotox Res 14(4):317–328

Ghosh S, May MJ, Kopp EB (1998) NF-kappa B and Rel proteins: evolutionarily conserved mediators of immune responses. Annu Rev Immunol 16:225–260

Ghoshal K, Jacob ST (2001) Regulation of metallothionein gene expression. Prog Nucleic Acid Res Mol Biol 66:357–384

Ghoshal K, Wang Y, Sheridan JF, Jacob ST (1998) Metallothionein induction in response to restraint stress. Transcriptional control, adaptation to stress, and role of glucocorticoid. J Biol Chem 273(43):27904–27910

Ghoshal K, Majumder S, Li Z, Bray TM, Jacob ST (1999) Transcriptional induction of metallothionein-I and -II genes in the livers of Cu, Zn-superoxide dismutase knockout mice. Biochem Biophys Res Commun 264(3):735–742

Giacomini CP, Leung SY, Chen X, Yuen ST, Kim YH, Bair E, Pollack JR (2005) A gene expression signature of genetic instability in colon cancer. Cancer Res 65(20):9200–9205

Gilmore TD (2006) Introduction to NF-kappaB: players, pathways, perspectives. Oncogene 25 (51):6680–6684

Gius D, Botero A, Shah S, Curry HA (1999) Intracellular oxidation/reduction status in the regulation of transcription factors NF-kappaB and AP-1. Toxicol Lett 106(2–3):93–106

Glennas A (1983) Gold resistance in cultured human cells possible role of metallothionein. Scand J Rheumatol Suppl 51:42–44

Gocke AR, Cravens PD, Ben LH, Hussain RZ, Northrop SC, Racke MK, Lovett-Racke AE (2007) T-bet regulates the fate of Th1 and Th17 lymphocytes in autoimmunity. J Immunol 178 (3):1341–1348

Goerdt S, Orfanos CE (1999) Other functions, other genes: alternative activation of antigen-presenting cells. Immunity 10(2):137–142

Goering PL, Klaassen CD (1983) Altered subcellular distribution of cadmium following cadmium pretreatment: possible mechanism of tolerance to cadmium-induced lethality. Toxicol Appl Pharmacol 70(2):195–203

Goering PL, Klaassen CD (1984) Resistance to cadmium-induced hepatotoxicity in immature rats. Toxicol Appl Pharmacol 74(3):321–329

Gomulkiewicz A, Podhorska-Okolow M, Szulc R, Smorag Z, Wojnar A, Zabel M, Dziegiel P (2010) Correlation between metallothionein (MT) expression and selected prognostic factors in ductal breast cancers. Folia Histochem Cytobiol 48(2):242–248

Gong YH, Elliott JL (2000) Metallothionein expression is altered in a transgenic murine model of familial amyotrophic lateral sclerosis. Exp Neurol 162(1):27–36

Goulding H, Jasani B, Pereira H, Reid A, Galea M, Bell JA, Elston CW, Robertson JF, Blamey RW, Nicholson RA et al (1995) Metallothionein expression in human breast cancer. Br J Cancer 72(4):968–972

Groten JP, Koeman JH, van Nesselrooij JH, Luten JB, Fentener van Vlissingen JM, Stenhuis WS, van Bladeren PJ (1994) Comparison of renal toxicity after long-term oral administration of cadmium chloride and cadmium-metallothionein in rats. Fundam Appl Toxicol 23(4):544–552

Gumulec J, Masarik M, Krizkova S, Adam V, Hubalek J, Hrabeta J, Eckschlager T, Stiborova M, Kizek R (2011) Insight to physiology and pathology of zinc(II) ions and their actions in breast and prostate carcinoma. Curr Med Chem 18(33):5041–5051

Gumulec J, Masarik M, Krizkova S, Hlavna M, Babula P, Hrabec R, Rovny A, Masarikova M, Sochor J, Adam V, Eckschlager T, Kizek R (2012) Evaluation of alpha-methylacyl-CoA racemase, metallothionein and prostate specific antigen as prostate cancer prognostic markers. Neoplasma 59(2):191–201

Gumulec J, Raudenska M, Adam V, Kizek R, Masarik M (2014) Metallothionein - immunohistochemical cancer biomarker: a meta-analysis. PLoS One 9(1), e85346

Gunes C, Heuchel R, Georgiev O, Muller KH, Lichtlen P, Bluthmann H, Marino S, Aguzzi A, Schaffner W (1998) Embryonic lethality and liver degeneration in mice lacking the metal-responsive transcriptional activator MTF-1. EMBO J 17(10):2846–2854

Guo R, Ma H, Gao F, Zhong L, Ren J (2009) Metallothionein alleviates oxidative stress-induced endoplasmic reticulum stress and myocardial dysfunction. J Mol Cell Cardiol 47(2):228–237

Gurel V, Sens DA, Somji S, Garrett SH, Nath J, Sens MA (2003) Stable transfection and overexpression of metallothionein isoform 3 inhibits the growth of MCF-7 and Hs578T cells but not that of T-47D or MDA-MB-231 cells. Breast Cancer Res Treat 80(2):181–191

Haase H, Rink L (2014) Zinc signals and immune function. Biofactors 40(1):27–40

Habeebu SS, Liu J, Liu Y, Klaassen CD (2000a) Metallothionein-null mice are more sensitive than wild-type mice to liver injury induced by repeated exposure to cadmium. Toxicol Sci 55(1):223–232

Habeebu SS, Liu J, Liu Y, Klaassen CD (2000b) Metallothionein-null mice are more susceptible than wild-type mice to chronic CdCl(2)-induced bone injury. Toxicol Sci 56(1):211–219

Habel N, Hamidouche Z, Girault I, Patiño-García A, Lecanda F, Marie PJ, Fromigué O (2013) Zinc chelation: a metallothionein 2A's mechanism of action involved in osteosarcoma cell death and chemotherapy resistance. Cell Death Dis 4, e874

Haerslev T, Jacobsen K, Nedergaard L, Zedeler K (1994) Immunohistochemical detection of metallothionein in primary breast carcinomas and their axillary lymph node metastases. Pathol Res Pract 190(7):675–681

Haerslev T, Jacobsen GK, Zedeler K (1995) The prognostic significance of immunohistochemically detectable metallothionein in primary breast carcinomas. APMIS 103(4):279–285

Hamza-Chaffai A, Amiard JC, Cosson RP (1999) Relationship between metallothioneins and metals in a natural population of the clam Ruditapes decussatus from Sfax coast: a non-linear model using Box-Cox transformation. Comp Biochem Physiol C Pharmacol Toxicol Endocrinol 123(2):153–163

Han YC, Zheng ZL, Zuo ZH, Yu YP, Chen R, Tseng GC, Nelson JB, Luo JH (2013) Metallothionein 1 h tumour suppressor activity in prostate cancer is mediated by euchromatin methyltransferase 1. J Pathol 230(2):184–193

Hanada K (2000) Photoprotective role of metallothionein in UV-injury - metallothionein-null mouse exhibits reduced tolerance against ultraviolet-B. J Dermatol Sci 23(Suppl 1):S51–56

Hanada K, Sawamura D, Hashimoto I, Kida K, Naganuma A (1998) Epidermal proliferation of the skin in metallothionein-null mice. J Invest Dermatol 110(3):259–262

Haq F, Mahoney M, Koropatnick J (2003) Signaling events for metallothionein induction. Mutat Res 533(1–2):211–226

Harley CB, Menon CR, Rachubinski RA, Nieboer E (1989) Metallothionein mRNA and protein induction by cadmium in peripheral-blood leucocytes. Biochem J 262(3):873–879

Hart BA, Voss GW, Willean CL (1989) Pulmonary tolerance to cadmium following cadmium aerosol pretreatment. Toxicol Appl Pharmacol 101(3):447–460

Hart BA, Gong Q, Eneman JD, Durieux-Lu CC (1995) In vivo expression of metallothionein in rat alveolar macrophages and type II epithelial cells following repeated cadmium aerosol exposures. Toxicol Appl Pharmacol 133(1):82–90

Hart BA, Potts RJ, Watkin RD (2001) Cadmium adaptation in the lung - a double-edged sword? Toxicology 160(1–3):65–70

Hartmann HJ, Gartner A, Weser U (1985) Copper dependent control of the enzymic and phagocyte induced degradation of some biopolymers, a possible link to systemic inflammation. Clin Chim Acta 152(1–2):95–103

Hartmann HJ, Schechinger T, Weser U (1989) Copper-thionein in leucocytes. Biol Met 2(1):40–44

Hashimoto K, Hayashi Y, Watabe K, Inuzuka T, Hozumi I (2011) Metallothionein-III prevents neuronal death and prolongs life span in amyotrophic lateral sclerosis model mice. Neuroscience 189:293–298

Hasumi M, Suzuki K, Matsui H, Koike H, Ito K, Yamanaka H (2003) Regulation of metallothionein and zinc transporter expression in human prostate cancer cells and tissues. Cancer Lett 200(2):187–195

Hatcher EL, Alexander JM, Kang YJ (1997) Decreased sensitivity to adriamycin in cadmium-resistant human lung carcinoma A549 cells. Biochem Pharmacol 53(5):747–754

Hayes RA, Regondi S, Winter MJ, Butler PJ, Agradi E, Taylor EW, Kevin Chipman J (2004) Cloning of a chub metallothionein cDNA and development of competitive RT-PCR of chub metallothionein mRNA as a potential biomarker of heavy metal exposure. Mar Environ Res 58 (2–5):665–669

He T, Wei D, Fabris D, Fenselau C (2000) Intracellular sequestration of anti-tumor drugs by metallothionein. Cell Mol Biol (Noisy-le-grand) 46(2):383–392

Hecht D, Jung D, Prabhu VV, Munson PJ, Hoffman MP, Kleinman HK (2002) Metallothionein promotes laminin-1-induced acinar differentiation in vitro and reduces tumor growth in vivo. Cancer Res 62(18):5370–5374

Hengstler JG, Pilch H, Schmidt M, Dahlenburg H, Sagemüller J, Schiffer I, Oesch F, Knapstein PG, Kaina B, Tanner B (2001) Metallothionein expression in ovarian cancer in relation to histopathological parameters and molecular markers of prognosis. Int J Cancer 95(2):121–127

Henkler F, Brinkmann J, Luch A (2010) The role of oxidative stress in carcinogenesis induced by metals and xenobiotics. Cancers (Basel) 2(2):376–396

Henrique R, Jerónimo C, Hoque MO, Nomoto S, Carvalho AL, Costa VL, Oliveira J, Teixeira MR, Lopes C, Sidransky D (2005) MT1G hypermethylation is associated with higher tumor stage in prostate cancer. Cancer Epidemiol Biomarkers Prev 14(5):1274–1278

Hernandez J, Carrasco J, Belloso E, Giralt M, Bluethmann H, Kee Lee D, Andrews GK, Hidalgo J (2000) Metallothionein induction by restraint stress: role of glucocorticoids and IL-6. Cytokine 12(6):791–796

Heuchel R, Radtke F, Georgiev O, Stark G, Aguet M, Schaffner W (1994) The transcription factor MTF-1 is essential for basal and heavy metal-induced metallothionein gene expression. EMBO J 13(12):2870–2875

Hidalgo J, Campmany L, Borras M, Garvey JS, Armario A (1988a) Metallothionein response to stress in rats: role in free radical scavenging. Am J Physiol 255(4 Pt 1):E518–524

Hidalgo J, Giralt M, Garvey JS, Armario A (1988b) Physiological role of glucocorticoids on rat serum and liver metallothionein in basal and stress conditions. Am J Physiol 254(1 Pt 1): E71–78

Hidalgo J, Dingman A, Garvey JS (1991) Role of extracellular zinc and copper on metallothionein regulation in cultured rat hepatocytes. Hepatology 14(4 Pt 1):648–654

Higashimoto M, Isoyama N, Ishibashi S, Inoue M, Takiguchi M, Suzuki S, Ohnishi Y, Sato M (2009) Tissue-dependent preventive effect of metallothionein against DNA damage in dyslipidemic mice under repeated stresses of fasting or restraint. Life Sci 84(17–18):569–575

Himeno S, Yanagiya T, Fujishiro H (2009) The role of zinc transporters in cadmium and manganese transport in mammalian cells. Biochimie 91(10):1218–1222

Hinkel A, Schmidtchen S, Palisaar RJ, Noldus J, Pannek J (2008) Identification of bladder cancer patients at risk for recurrence or progression: an immunohistochemical study based on the expression of metallothionein. J Toxicol Environ Health A 71(13–14):954–959

Hishikawa Y, Abe S, Kinugasa S, Yoshimura H, Monden N, Igarashi M, Tachibana M, Nagasue N (1997) Overexpression of metallothionein correlates with chemoresistance to cisplatin and prognosis in esophageal cancer. Oncology 54(4):342–347

Hishikawa Y, Koji T, Dhar DK, Kinugasa S, Yamaguchi M, Nagasue N (1999) Metallothionein expression correlates with metastatic and proliferative potential in squamous cell carcinoma of the oesophagus. Br J Cancer 81(4):712–720

Hishikawa Y, Kohno H, Ueda S, Kimoto T, Dhar DK, Kubota H, Tachibana M, Koji T, Nagasue N (2001) Expression of metallothionein in colorectal cancers and synchronous liver metastases. Oncology 61(2):162–167

Hiura T, Khalid H, Yamashita H, Tokunaga Y, Yasunaga A, Shibata S (1998) Immunohistochemical analysis of metallothionein in astrocytic tumors in relation to tumor grade, proliferative potential, and survival. Cancer 83(11):2361–2369

Hoey JG, Garrett SH, Sens MA, Todd JH, Sens DA (1997) Expression of MT-3 mRNA in human kidney, proximal tubule cell cultures, and renal cell carcinoma. Toxicol Lett 92(2):149–160

Hoogenraad TU (2006) Paradigm shift in treatment of Wilson's disease: zinc therapy now treatment of choice. Brain Dev 28(3):141–146

Howells C, West AK, Chung RS (2010) Neuronal growth-inhibitory factor (metallothionein-3): evaluation of the biological function of growth-inhibitory factor in the injured and neurodegenerative brain. FEBS J 277(14):2931–2939

Hozumi I (2013) Roles and therapeutic potential of metallothioneins in neurodegenerative diseases. Curr Pharm Biotechnol 14(4):408–413

Hozumi I, Inuzuka T, Tsuji S (1998) Brain injury and growth inhibitory factor (GIF)--a minireview. Neurochem Res 23(3):319–328

Hozumi I, Suzuki JS, Kanazawa H, Hara A, Saio M, Inuzuka T, Miyairi S, Naganuma A, Tohyama C (2008) Metallothionein-3 is expressed in the brain and various peripheral organs of the rat. Neurosci Lett 438(1):54–58

Hsieh HL, Yang CM (2013) Role of redox signaling in neuroinflammation and neurodegenerative diseases. Biomed Res Int 2013:484613

Hu N, Han X, Lane EK, Gao F, Zhang Y, Ren J (2013) Cardiac-specific overexpression of metallothionein rescues against cigarette smoking exposure-induced myocardial contractile and mitochondrial damage. PLoS One 8(2), e57151

Huang IY, Yoshida A (1977) Mouse liver metallothioneins. Complete amino acid sequence of metallothionein-I. J Biol Chem 252(22):8217–8221

Huang Y, Prasad M, Lemon WJ, Hampel H, Wright FA, Kornacker K, LiVolsi V, Frankel W, Kloos RT, Eng C, Pellegata NS, de la Chapelle A (2001) Gene expression in papillary thyroid carcinoma reveals highly consistent profiles. Proc Natl Acad Sci USA 98(26):15044–15049

Huang Y, de la Chapelle A, Pellegata NS (2003) Hypermethylation, but not LOH, is associated with the low expression of MT1G and CRABP1 in papillary thyroid carcinoma. Int J Cancer 104(6):735–744

Huang M, Shaw IC, Petering DH (2004) Interprotein metal exchange between transcription factor IIIa and apo-metallothionein. J Inorg Biochem 98(4):639–648

Huber KL, Cousins RJ (1993) Metallothionein expression in rat bone marrow is dependent on dietary zinc but not dependent on interleukin-1 or interleukin-6. J Nutr 123(4):642–648

Huh S, Lee K, Yun HS, Paik DJ, Kim JM, Youn J (2007) Functions of metallothionein generating interleukin-10-producing regulatory CD4+ T cells potentiate suppression of collagen-induced arthritis. J Microbiol Biotechnol 17(2):348–358

Hwang TL, Chen HY, Changchien TT, Wang CC, Wu CM (2013) The cytotoxicity of mercury chloride to the keratinocytes is associated with metallothionein expression. Biomed Rep 1(3):379–382

Ichibangase Y, Yamamoto M, Yasuda M, Houki N, Nobunaga M (1998) Induction of drug resistance to gold sodium thiomalate in a monocyte cell line, THP-1. Clin Rheumatol 17(3):214–218

Ilback NG, Glynn AW, Wikberg L, Netzel E, Lindh U (2004) Metallothionein is induced and trace element balance changed in target organs of a common viral infection. Toxicology 199(2–3):241–250

Imoto AM, Okada M, Okazaki T, Kitasato H, Harigae H, Takahashi S (2010) Metallothionein-1 isoforms and vimentin are direct PU.1 downstream target genes in leukemia cells. J Biol Chem 285(14):10300–10309

Inomata M, Takahashi S, Harigae H, Kameoka J, Kaku M, Sasaki T (2006) Inverse correlation between Flt3 and PU.1 expression in acute myeloblastic leukemias. Leuk Res 30(6):659–664

Inoue K, Takano H, Shimada A, Wada E, Yanagisawa R, Sakurai M, Satoh M, Yoshikawa T (2006) Role of metallothionein in coagulatory disturbance and systemic inflammation induced by lipopolysaccharide in mice. FASEB J 20(3):533–535

Inoue K, Takano H, Kaewamatawong T, Shimada A, Suzuki J, Yanagisawa R, Tasaka S, Ishizaka A, Satoh M (2008) Role of metallothionein in lung inflammation induced by ozone exposure in mice. Free Radic Biol Med 45(12):1714–1722

Ioachim EE, Assimakopoulos D, Peschos D, Zissi A, Skevas A, Agnantis NJ (1999) Immunohistochemical expression of metallothionein in benign premalignant and malignant epithelium of the larynx: correlation with p53 and proliferative cell nuclear antigen. Pathol Res Pract 195(12):809–814

Ioachim EE, Kitsiou E, Carassavoglou C, Stefanaki S, Agnantis NJ (2000) Immunohistochemical localization of metallothionein in endometrial lesions. J Pathol 191(3):269–273

Ioachim EE, Charchanti AV, Stavropoulos NE, Athanassiou ED, Michael MC, Agnantis NJ (2001) Localization of metallothionein in urothelial carcinoma of the human urinary bladder: an immunohistochemical study including correlation with HLA-DR antigen, p53, and proliferation indices. Anticancer Res 21(3B):1757–1761

Ioachim EE, Tsanou E, Briasoulis E, Batsis C, Karavasilis V, Charchanti A, Pavlidis N, Agnantis NJ (2003) Clinicopathological study of the expression of hsp27, pS2, cathepsin D and metallothionein in primary invasive breast cancer. Breast 12(2):111–119

Ishii K, Usui S, Yamamoto H, Sugimura Y, Tatematsu M, Hirano K (2001) Decreases of metallothionein and aminopeptidase N in renal cancer tissues. J Biochem 129(2):253–258

Iwata M, Takebayashi T, Ohta H, Alcalde RE, Itano Y, Matsumura T (1999) Zinc accumulation and metallothionein gene expression in the proliferating epidermis during wound healing in mouse skin. Histochem Cell Biol 112(4):283–290

Izawa JI, Moussa M, Cherian MG, Doig G, Chin JL (1998) Metallothionein expression in renal cancer. Urology 52(5):767–772

Jacob ST, Ghoshal K, Sheridan JF (1999) Induction of metallothionein by stress and its molecular mechanisms. Gene Expr 7(4–6):301–310

Janssen AM, van Duijn W, Oostendorp-Van De Ruit MM, Kruidenier L, Bosman CB, Griffioen G, Lamers CB, van Krieken JH, van De Velde CJ, Verspaget HW (2000) Metallothionein in human gastrointestinal cancer. J Pathol 192(3):293–300

Janssen AM, van Duijn W, Kubben FJ, Griffioen G, Lamers CB, van Krieken JH, van de Velde CJ, Verspaget HW (2002) Prognostic significance of metallothionein in human gastrointestinal cancer. Clin Cancer Res 8(6):1889–1896

Jia G, Gu YQ, Chen KT, Lu YY, Yan L, Wang JL, Su YP, Wu JC (2004a) Protective role of metallothionein (I/II) against pathological damage and apoptosis induced by dimethylarsinic acid. World J Gastroenterol 10(1):91–95

Jia G, Sone H, Nishimura N, Satoh M, Tohyama C (2004b) Metallothionein (I/II) suppresses genotoxicity caused by dimethylarsinic acid. Int J Oncol 25(2):325–333

Jin R, Bay BH, Chow VT, Tan PH, Lin VC (2000) Metallothionein 1E mRNA is highly expressed in oestrogen receptor-negative human invasive ductal breast cancer. Br J Cancer 83(3):319–323

Jin R, Bay BH, Chow VT, Tan PH (2001) Metallothionein 1F mRNA expression correlates with histological grade in breast carcinoma. Breast Cancer Res Treat 66(3):265–272

Jin R, Chow VT, Tan PH, Dheen ST, Duan W, Bay BH (2002) Metallothionein 2A expression is associated with cell proliferation in breast cancer. Carcinogenesis 23(1):81–86

Johann AC, da Silveira-Junior JB, Souto GR, Horta MC, Aguiar MC, Mesquita RA (2008) Metallothionein immunoexpression in oral leukoplakia. Med Oral Patol Oral Cir Bucal 13(3):E156–160

Joseph MG, Banerjee D, Kocha W, Feld R, Stitt LW, Cherian MG (2001) Metallothionein expression in patients with small cell carcinoma of the lung: correlation with other molecular markers and clinical outcome. Cancer 92(4):836–842

Juang HH, Chung LC, Sung HC, Feng TH, Lee YH, Chang PL, Tsui KH (2013) Metallothionein 3: an androgen-upregulated gene enhances cell invasion and tumorigenesis of prostate carcinoma cells. Prostate 73(14):1495–1506

Kanekiyo M, Itoh N, Kawasaki A, Tanaka J, Nakanishi T, Tanaka K (2001) Zinc-induced activation of the human cytomegalovirus major immediate-early promoter is mediated by metallothionein and nuclear factor-kappaB. Toxicol Appl Pharmacol 173(3):146–153

Karasawa M, Nishimura N, Nishimura H, Tohyama C, Hashiba H, Kuroki T (1991) Localization of metallothionein in hair follicles of normal skin and the basal cell layer of hyperplastic epidermis: possible association with cell proliferation. J Invest Dermatol 97(1):97–100

Karin M, Herschman HR (1980) Glucocorticoid hormone receptor mediated induction of metallothionein synthesis in HeLa cells. J Cell Physiol 103(1):35–40

Karin M, Eddy RL, Henry WM, Haley LL, Byers MG, Shows TB (1984) Human metallothionein genes are clustered on chromosome 16. Proc Natl Acad Sci USA 81(17):5494–5498

Karotki AV, Vasak M (2009) Reaction of human metallothionein-3 with cisplatin and transplatin. J Biol Inorg Chem 14(7):1129–1138

Kawashima T, Doh-ura K, Torisu M, Uchida Y, Furuta A, Iwaki T (2000) Differential expression of metallothioneins in human prion diseases. Dement Geriatr Cogn Disord 11(5):251–262

Kelley SL, Basu A, Teicher BA, Hacker MP, Hamer DH, Lazo JS (1988) Overexpression of metallothionein confers resistance to anticancer drugs. Science 241(4874):1813–1815

Kelly EJ, Sandgren EP, Brinster RL, Palmiter RD (1997) A pair of adjacent glucocorticoid response elements regulate expression of two mouse metallothionein genes. Proc Natl Acad Sci USA 94(19):10045–10050

Kenaga C, Cherian MG, Cox C, Oberdorster G (1996) Metallothionein induction and pulmonary responses to inhaled cadmium chloride in rats and mice. Fundam Appl Toxicol 30(2):204–212

Kennette W, Collins OM, Zalups RK, Koropatnick J (2005) Basal and zinc-induced metallothionein in resistance to cadmium, cisplatin, zinc, and tertbutyl hydroperoxide: studies using MT knockout and antisense-downregulated MT in mammalian cells. Toxicol Sci 88 (2):602–613

Khan KH, Blanco-Codesido M, Molife LR (2014) Cancer therapeutics: Targeting the apoptotic pathway. Crit Rev Oncol Hematol 90(3):200–219

Kim SJ, Diamond B (2015) Modulation of tolerogenic dendritic cells and autoimmunity. Semin Cell Dev Biol 41:49–58

Kim CH, Kim JH, Hsu CY, Ahn YS (1999) Zinc is required in pyrrolidine dithiocarbamate inhibition of NF-kappaB activation. FEBS Lett 449(1):28–32

Kim CH, Kim JH, Lee J, Ahn YS (2003) Zinc-induced NF-kappaB inhibition can be modulated by changes in the intracellular metallothionein level. Toxicol Appl Pharmacol 190(2):189–196

Kim HG, Kim JY, Han EH, Hwang YP, Choi JH, Park BH, Jeong HG (2011) Metallothionein-2A overexpression increases the expression of matrix metalloproteinase-9 and invasion of breast cancer cells. FEBS Lett 585(2):421–428

Kimura T, Itoh N, Takehara M, Oguro I, Ishizaki JI, Nakanishi T, Tanaka K (2001) Sensitivity of metallothionein-null mice to LPS/D-galactosamine-induced lethality. Biochem Biophys Res Commun 280(1):358–362

Klasing KC (1984) Effect of inflammatory agents and interleukin 1 on iron and zinc metabolism. Am J Physiol 247(5 Pt 2):R901–904

Klassen RB, Crenshaw K, Kozyraki R, Verroust PJ, Tio L, Atrian S, Allen PL, Hammond TG (2004) Megalin mediates renal uptake of heavy metal metallothionein complexes. Am J Physiol Renal Physiol 287(3):F393–403

Klein D, Lichtmannegger J, Heinzmann U, Muller-Hocker J, Michaelsen S, Summer KH (1998) Association of copper to metallothionein in hepatic lysosomes of Long-Evans cinnamon (LEC) rats during the development of hepatitis. Eur J Clin Invest 28(4):302–310

Klosterhalfen B, Tons C, Hauptmann S, Tietze L, Offner FA, Kupper W, Kirkpatrick CJ (1996) Influence of heat shock protein 70 and metallothionein induction by zinc-bis-(DL-hydrogen-aspartate) on the release of inflammatory mediators in a porcine model of recurrent endotoxemia. Biochem Pharmacol 52(8):1201–1210

Kmiecik AM, Pula B, Suchanski J, Olbromski M, Gomulkiewicz A, Owczarek T, Kruczak A, Ambicka A, Rys J, Ugorski M, Podhorska-Okolow M, Dziegiel P (2015) Metallothionein-3 Increases Triple-Negative Breast Cancer Cell Invasiveness via Induction of Metalloproteinase Expression. PLoS One 10(5), e0124865

Kobayashi H, Uchida Y, Ihara Y, Nakajima K, Kohsaka S, Miyatake T, Tsuji S (1993) Molecular cloning of rat growth inhibitory factor cDNA and the expression in the central nervous system. Brain Res Mol Brain Res 19(3):188–194

Kobayashi K, Kuroda J, Shibata N, Hasegawa T, Seko Y, Satoh M, Tohyama C, Takano H, Imura N, Sakabe K, Fujishiro H, Himeno S (2007) Induction of metallothionein by manganese is completely dependent on interleukin-6 production. J Pharmacol Exp Ther 320(2):721–727

Kobierzycki C, Pula B, Skiba M, Jablonska K, Latkowski K, Zabel M, Nowak-Markwitz E, Spaczynski M, Kedzia W, Podhorska-Okolow M, Dziegiel P (2013) Comparison of minichromosome maintenance proteins (MCM-3, MCM-7) and metallothioneins (MT-I/II, MT-III) expression in relation to clinicopathological data in ovarian cancer. Anticancer Res 33(12):5375–5383

Koizumi S, Suzuki K, Ogra Y, Yamada H, Otsuka F (1999) Transcriptional activity and regulatory protein binding of metal-responsive elements of the human metallothionein-IIA gene. Eur J Biochem 259(3):635–642

Kojima Y, Berger C, Vallee BL, Kagi JH (1976) Amino-acid sequence of equine renal metallothionein-1B. Proc Natl Acad Sci USA 73(10):3413–3417

Kojima I, Tanaka T, Inagi R, Nishi H, Aburatani H, Kato H, Miyata T, Fujita T, Nangaku M (2009) Metallothionein is upregulated by hypoxia and stabilizes hypoxia-inducible factor in the kidney. Kidney Int 75(3):268–277

Kondo Y, Kuo SM, Watkins SC, Lazo JS (1995a) Metallothionein localization and cisplatin resistance in human hormone-independent prostatic tumor cell lines. Cancer Res 55 (3):474–477

Kondo Y, Woo ES, Michalska AE, Choo KH, Lazo JS (1995b) Metallothionein null cells have increased sensitivity to anticancer drugs. Cancer Res 55(10):2021–2023

Kondo Y, Rusnak JM, Hoyt DG, Settineri CE, Pitt BR, Lazo JS (1997) Enhanced apoptosis in metallothionein null cells. Mol Pharmacol 52(2):195–201

Kondo Y, Himeno S, Endo W, Mita M, Suzuki Y, Nemoto K, Akimoto M, Lazo JS, Imura N (1999) Metallothionein modulates the carcinogenicity of N-butyl-N-(4-hydroxybutyl)nitrosa-mine in mice. Carcinogenesis 20(8):1625–1627

Kondo Y, Yamagata K, Satoh M, Himeno S, Imura N, Nishimura T (2003) Optimal administration schedule of cisplatin for bladder tumor with minimal induction of metallothionein.". J Urol 170 (6 Pt 1):2467–2470

Koropatnick J, Zalups RK (1997) Effect of non-toxic mercury, zinc or cadmium pretreatment on the capacity of human monocytes to undergo lipopolysaccharide-induced activation. Br J Pharmacol 120(5):797–806

Korshunov A, Sycheva R, Timirgaz V, Golanov A (1999) Prognostic value of immunoexpression of the chemoresistance-related proteins in ependymomas: an analysis of 76 cases. J Neurooncol 45(3):219–227

Kramer KK, Zoelle JT, Klaassen CD (1996) Induction of metallothionein mRNA and protein in primary murine neuron cultures. Toxicol Appl Pharmacol 141(1):1–7

Krepkiy D, Antholine WE, Petering DH (2003) Properties of the reaction of chromate with metallothionein. Chem Res Toxicol 16(6):750–756

Kreppel H, Bauman JW, Liu J, McKim JM Jr, Klaassen CD (1993) Induction of metallothionein by arsenicals in mice. Fundam Appl Toxicol 20(2):184–189

Krizkova S, Adam V, Kizek R (2009a) Study of metallothionein oxidation by using of chip CE. Electrophoresis 30(23):4029–4033

Krizkova S, Fabrik I, Adam V, Hrabeta J, Eckschlager T, Kizek R (2009b) Metallothionein--a promising tool for cancer diagnostics. Bratisl Lek Listy 110(2):93–97

Krizkova S, Masarik M, Majzlik P, Kukacka J, Kruseova J, Adam V, Prusa R, Eckschlager T, Stiborova M, Kizek R (2010) Serum metallothionein in newly diagnosed patients with childhood solid tumours. Acta Biochim Pol 57(4):561–566

Krizkova S, Ryvolova M, Hrabeta J, Adam V, Stiborova M, Eckschlager T, Kizek R (2012) Metallothioneins and zinc in cancer diagnosis and therapy. Drug Metab Rev 44(4):287–301

Krolicka A, Kobierzycki C, Pula B, Podhorska-Okolow M, Piotrowska A, Rzeszutko M, Rzeszutko W, Rabczynski J, Domoslawski P, Wojtczak B, Dawiskiba J, Dziegiel P (2010) Comparison of metallothionein (MT) and Ki-67 antigen expression in benign and malignant thyroid tumours. Anticancer Res 30(12):4945–4949

Kruseova J, Hynek D, Adam V, Kizek R, Prusa R, Hrabeta J, Eckschlager T (2013) Serum metallothioneins in childhood tumours-a potential prognostic marker. Int J Mol Sci 14 (6):12170–12185

Krześlak A, Forma E, Chwatko G, Jóźwiak P, Szymczyk A, Wilkosz J, Różański W, Bryś M (2013) Effect of metallothionein 2A gene polymorphism on allele-specific gene expression and metal content in prostate cancer. Toxicol Appl Pharmacol 268(3):278–285

Krzeslak A, Forma E, Jozwiak P, Szymczyk A, Smolarz B, Romanowicz-Makowska H, Rozanski W, Brys M (2014) Metallothionein 2A genetic polymorphisms and risk of ductal breast cancer. Clin Exp Med 14(1):107–113

Kuchenbauer F, Kern W, Schoch C, Kohlmann A, Hiddemann W, Haferlach T, Schnittger S (2005) Detailed analysis of FLT3 expression levels in acute myeloid leukemia. Haematologica 90(12):1617–1625

Kumar P, Lal NR, Mondal AK, Mondal A, Gharami RC, Maiti A (2012) Zinc and skin: a brief summary. Dermatol Online J 18(3):1

Lai Y, Lim D, Tan PH, Leung TK, Yip GW, Bay BH (2010) Silencing the Metallothionein-2A gene induces entosis in adherent MCF-7 breast cancer cells. Anat Rec (Hoboken) 293 (10):1685–1691

Langmade SJ, Ravindra R, Daniels PJ, Andrews GK (2000) The transcription factor MTF-1 mediates metal regulation of the mouse ZnT1 gene. J Biol Chem 275(44):34803–34809

Lee SJ, Koh JY (2010) Roles of zinc and metallothionein-3 in oxidative stress-induced lysosomal dysfunction, cell death, and autophagy in neurons and astrocytes. Mol Brain 3(1):30

Lee DK, Carrasco J, Hidalgo J, Andrews GK (1999) Identification of a signal transducer and activator of transcription (STAT) binding site in the mouse metallothionein-I promoter involved in interleukin-6-induced gene expression. Biochem J 337(Pt 1):59–65

Lee SS, Yang SF, Ho YC, Tsai CH, Chang YC (2008) The upregulation of metallothionein-1 expression in areca quid chewing-associated oral squamous cell carcinomas. Oral Oncol 44 (2):180–186

Lee JD, Wu SM, Lu LY, Yang YT, Jeng SY (2009) Cadmium concentration and metallothionein expression in prostate cancer and benign prostatic hyperplasia of humans. J Formos Med Assoc 108(7):554–559

Lee SJ, Park MH, Kim HJ, Koh JY (2010) Metallothionein-3 regulates lysosomal function in cultured astrocytes under both normal and oxidative conditions. Glia 58(10):1186–1196

Lee SJ, Cho KS, Kim HN, Kim HJ, Koh JY (2011) Role of zinc metallothionein-3 (ZnMt3) in epidermal growth factor (EGF)-induced c-Abl protein activation and actin polymerization in cultured astrocytes. J Biol Chem 286(47):40847–40856

Lee SJ, Seo BR, Choi EJ, Koh JY (2014) The role of reciprocal activation of cAbl and Mst1 in the oxidative death of cultured astrocytes. Glia 62(4):639–648

Leibbrandt ME, Khokha R, Koropatnick J (1994) Antisense down-regulation of metallothionein in a human monocytic cell line alters adherence, invasion, and the respiratory burst. Cell Growth Differ 5(1):17–25

Levadoux-Martin M, Hesketh JE, Beattie JH, Wallace HM (2001) Influence of metallothionein-1 localization on its function. Biochem J 355(Pt 2):473–479

Levings MK, Roncarolo MG (2000) T-regulatory 1 cells: a novel subset of CD4 T cells with immunoregulatory properties. J Allergy Clin Immunol 106(1 Pt 2):S109–112

Lewinski A, Sliwka PW, Stasiolek M (2014) Dendritic cells in autoimmune disorders and cancer of the thyroid. Folia Histochem Cytobiol 52(1):18–28

Li Y, Wo JM, Cai L, Zhou Z, Rosenbaum D, Mendez C, Ray MB, Jones WF, Kang YJ (2003) Association of metallothionein expression and lack of apoptosis with progression of carcinogenesis in Barrett's esophagus. Exp Biol Med (Maywood) 228(3):286–292

Li X, Chen H, Epstein PN (2004) Metallothionein protects islets from hypoxia and extends islet graft survival by scavenging most kinds of reactive oxygen species. J Biol Chem 279 (1):765–771

Liang L, Fu K, Lee DK, Sobieski RJ, Dalton T, Andrews GK (1996) Activation of the complete mouse metallothionein gene locus in the maternal deciduum. Mol Reprod Dev 43(1):25–37

Lichtlen P, Schaffner W (2001) The "metal transcription factor" MTF-1: biological facts and medical implications. Swiss Med Wkly 131(45–46):647–652

Lim D, Jocelyn KM, Yip GW, Bay BH (2009) Silencing the Metallothionein-2A gene inhibits cell cycle progression from G1- to S-phase involving ATM and cdc25A signaling in breast cancer cells. Cancer Lett 276(1):109–117

Lin SF, Wei H, Maeder D, Franklin RB, Feng P (2009) Profiling of zinc-altered gene expression in human prostate normal vs. cancer cells: a time course study. J Nutr Biochem 20 (12):1000–1012

Liu J, Liu Y, Klaassen CD (1994) Nephrotoxicity of CdCl2 and Cd-metallothionein in cultured rat kidney proximal tubules and LLC-PK1 cells. Toxicol Appl Pharmacol 128(2):264–270

Liu Y, Liu J, Iszard MB, Andrews GK, Palmiter RD, Klaassen CD (1995) Transgenic mice that overexpress metallothionein-I are protected from cadmium lethality and hepatotoxicity. Toxicol Appl Pharmacol 135(2):222–228

Liu J, Liu Y, Michalska AE, Choo KH, Klaassen CD (1996) Metallothionein plays less of a protective role in cadmium-metallothionein-induced nephrotoxicity than in cadmium chloride-induced hepatotoxicity. J Pharmacol Exp Ther 276(3):1216–1223

Liu J, Liu Y, Habeebu SS, Klaassen CD (1998) Susceptibility of MT-null mice to chronic CdCl2-induced nephrotoxicity indicates that renal injury is not mediated by the CdMT complex. Toxicol Sci 46(1):197–203

Liu J, Liu Y, Habeebu SS, Klaassen CD (1999) Metallothionein-null mice are highly susceptible to the hematotoxic and immunotoxic effects of chronic CdCl2 exposure. Toxicol Appl Pharmacol 159(2):98–108

Liu J, Liu Y, Goyer RA, Achanzar W, Waalkes MP (2000a) Metallothionein-I/II null mice are more sensitive than wild-type mice to the hepatotoxic and nephrotoxic effects of chronic oral or injected inorganic arsenicals. Toxicol Sci 55(2):460–467

Liu J, Liu Y, Habeebu SM, Waalkes MP, Klaassen CD (2000b) Chronic combined exposure to cadmium and arsenic exacerbates nephrotoxicity, particularly in metallothionein-I/II null mice. Toxicology 147(3):157–166

Liu CG, Zhang L, Jiang Y, Chatterjee D, Croce CM, Huebner K, Fong LY (2005) Modulation of gene expression in precancerous rat esophagus by dietary zinc deficit and replenishment. Cancer Res 65(17):7790–7799

Liu ZM, Chen GG, Shum CK, Vlantis AC, Cherian MG, Koropatnick J, van Hasselt CA (2007) Induction of functional MT1 and MT2 isoforms by calcium in anaplastic thyroid carcinoma cells. FEBS Lett 581(13):2465–2472

Liu ZM, Hasselt CA, Song FZ, Vlantis AC, Cherian MG, Koropatnick J, Chen GG (2009) Expression of functional metallothionein isoforms in papillary thyroid cancer. Mol Cell Endocrinol 302(1):92–98

Lock C, Hermans G, Pedotti R, Brendolan A, Schadt E, Garren H, Langer-Gould A, Strober S, Cannella B, Allard J, Klonowski P, Austin A, Lad N, Kaminski N, Galli SJ, Oksenberg JR,

Raine CS, Heller R, Steinman L (2002) Gene-microarray analysis of multiple sclerosis lesions yields new targets validated in autoimmune encephalomyelitis. Nat Med 8(5):500–508

Luo Y, Xu Y, Bao Q, Ding Z, Zhu C, Huang ZX, Tan X (2013) The molecular mechanism for human metallothionein-3 to protect against the neuronal cytotoxicity of Abeta(1-42) with Cu ions. J Biol Inorg Chem 18(1):39–47

Lynes MA, Garvey JS, Lawrence DA (1990) Extracellular metallothionein effects on lymphocyte activities. Mol Immunol 27(3):211–219

Lynes MA, Borghesi LA, Youn J, Olson EA (1993) Immunomodulatory activities of extracellular metallothionein. I Metallothionein effects on antibody production. Toxicology 85 (2–3):161–177

Lynes MA, Richardson CA, McCabe R, Crowthers KC, Lee JC, Youn J, Schweitzer IB, Shultz LD (1999) Metallothionein-mediated alterations in autoimmune disease processes. Klaassen CD. Basel, Birkahuser, pp 437–444

Lynn NN, Howe MC, Hale RJ, Collins GN, O'Reilly PH (2003) Over expression of metallothionein predicts resistance of transitional cell carcinoma of bladder to intravesical mitomycin therapy. J Urol 169(2):721–723

Ma C, Li LF, Chen X (2011) Expression of metallothionein-I and II in skin ageing and its association with skin proliferation. Br J Dermatol 164(3):479–482

MacDonald RS (2000) The role of zinc in growth and cell proliferation. J Nutr 130 (5S Suppl):1500S–1508S

Maghdooni Bagheri P, De Ley M (2011) Metallothionein in human immunomagnetically selected CD34(+) haematopoietic progenitor cells. Cell Biol Int 35(1):39–44

Maier H, Jones C, Jasani B, Ofner D, Zelger B, Schmid KW, Budka H (1997) Metallothionein overexpression in human brain tumours. Acta Neuropathol 94(6):599–604

Majumder S, Roy S, Kaffenberger T, Wang B, Costinean S, Frankel W, Bratasz A, Kuppusamy P, Hai T, Ghoshal K, Jacob ST (2010) Loss of metallothionein predisposes mice to diethylnitrosamine-induced hepatocarcinogenesis by activating NF-kappaB target genes. Cancer Res 70(24):10265–10276

Mannick EE, Schurr JR, Zapata A, Lentz JJ, Gastanaduy M, Cote RL, Delgado A, Correa P, Correa H (2004) Gene expression in gastric biopsies from patients infected with Helicobacter pylori. Scand J Gastroenterol 39(12):1192–1200

Manso Y, Carrasco J, Comes G, Adlard PA, Bush AI, Hidalgo J (2012) Characterization of the role of the antioxidant proteins metallothioneins 1 and 2 in an animal model of Alzheimer's disease. Cell Mol Life Sci 69(21):3665–3681

Mao P, Hever MP, Niemaszyk LM, Haghkerdar JM, Yanco EG, Desai D, Beyrouthy MJ, Kerley-Hamilton JS, Freemantle SJ, Spinella MJ (2011) Serine/threonine kinase 17A is a novel p53 target gene and modulator of cisplatin toxicity and reactive oxygen species in testicular cancer cells. J Biol Chem 286(22):19381–19391

Mao J, Yu H, Wang C, Sun L, Jiang W, Zhang P, Xiao Q, Han D, Saiyin H, Zhu J, Chen T, Roberts LR, Huang H, Yu L (2012) Metallothionein MT1M is a tumor suppressor of human hepatocellular carcinomas. Carcinogenesis 33(12):2568–2577

Maret W, Vallee BL (1998) Thiolate ligands in metallothionein confer redox activity on zinc clusters. Proc Natl Acad Sci USA 95(7):3478–3482

Maret W, Larsen KS, Vallee BL (1997) Coordination dynamics of biological zinc "clusters" in metallothioneins and in the DNA-binding domain of the transcription factor Gal4. Proc Natl Acad Sci USA 94(6):2233–2237

Margoshes M, Vallee B (1957) A cadmium protein from equine kidney cortex. J Am Chem Soc 79:1813–1814

Martinez FO, Helming L, Gordon S (2009) Alternative activation of macrophages: an immunologic functional perspective. Annu Rev Immunol 27:451–483

Masson-Lecomte A, Rava M, Real FX, Hartmann A, Allory Y, Malats N (2014) Inflammatory biomarkers and bladder cancer prognosis: a systematic review. Eur Urol 66(6):1078–1091

Masters BA, Quaife CJ, Erickson JC, Kelly EJ, Froelick GJ, Zambrowicz BP, Brinster RL, Palmiter RD (1994) Metallothionein III is expressed in neurons that sequester zinc in synaptic vesicles. J Neurosci 14(10):5844–5857

Mayer F, Stoop H, Scheffer GL, Scheper R, Oosterhuis JW, Looijenga LH, Bokemeyer C (2003) Molecular determinants of treatment response in human germ cell tumors. Clin Cancer Res 9 (2):767–773

McCluggage WG, Maxwell P, Bharucha H (1998) Immunohistochemical detection of metallothionein and MIB1 in uterine cervical squamous lesions. Int J Gynecol Pathol 17 (1):29–35

McCluggage WG, Maxwell P, Hamilton PW, Jasani B (1999) High metallothionein expression is associated with features predictive of aggressive behaviour in endometrial carcinoma. Histopathology 34(1):51–55

McCluggage WG, Strand K, Abdulkadir A (2002) Immunohistochemical localization of metallothionein in benign and malignant epithelial ovarian tumors. Int J Gynecol Cancer 12 (1):62–65

McGee HM, Woods GM, Bennett B, Chung RS (2010) The two faces of metallothionein in carcinogenesis: photoprotection against UVR-induced cancer and promotion of tumour survival. Photochem Photobiol Sci 9(4):586–596

McKim JM Jr, Choudhuri S, Klaassen CD (1992) In vitro degradation of apo-, zinc-, and cadmium-metallothionein by cathepsins B, C, and D. Toxicol Appl Pharmacol 116(1):117–124

Meijer C, Timmer A, De Vries EG, Groten JP, Knol A, Zwart N, Dam WA, Sleijfer DT, Mulder NH (2000) Role of metallothionein in cisplatin sensitivity of germ-cell tumours. Int J Cancer 85(6):777–781

Meloni G, Zovo K, Kazantseva J, Palumaa P, Vasak M (2006) Organization and assembly of metal-thiolate clusters in epithelium-specific metallothionein-4. J Biol Chem 281 (21):14588–14595

Meplan C, Verhaegh G, Richard MJ, Hainaut P (1999) Metal ions as regulators of the conformation and function of the tumour suppressor protein p53: implications for carcinogenesis. Proc Nutr Soc 58(3):565–571

Meplan C, Richard MJ, Hainaut P (2000) Redox signalling and transition metals in the control of the p53 pathway. Biochem Pharmacol 59(1):25–33

Miaw SC, Choi A, Yu E, Kishikawa H, Ho IC (2000) ROG, repressor of GATA, regulates the expression of cytokine genes. Immunity 12(3):323–333

Michael GJ, Esmailzadeh S, Moran LB, Christian L, Pearce RK, Graeber MB (2011) Up-regulation of metallothionein gene expression in parkinsonian astrocytes. Neurogenetics 12(4):295–305

Mididoddi S, McGuirt JP, Sens MA, Todd JH, Sens DA (1996) Isoform-specific expression of metallothionein mRNA in the developing and adult human kidney. Toxicol Lett 85(1):17–27

Miesel R, Zuber M (1993) Copper-dependent antioxidase defenses in inflammatory and autoimmune rheumatic diseases. Inflammation 17(3):283–294

Miesel R, Hartmann HJ, Weser U (1990) Antiinflammatory reactivity of copper(I)-thionein. Inflammation 14(5):471–483

Milnerowicz H, Chmarek M, Rabczynski J, Milnerowicz S, Nabzdyk S, Knast W (2004) Immunohistochemical localization of metallothionein in chronic pancreatitis. Pancreas 29(1):28–32

Milnerowicz H, Jablonowska M, Bizon A (2009) Change of zinc, copper, and metallothionein concentrations and the copper-zinc superoxide dismutase activity in patients with pancreatitis. Pancreas 38(6):681–688

Min KS, Mukai S, Ohta M, Onosaka S, Tanaka K (1992) Glucocorticoid inhibition of inflammation-induced metallothionein synthesis in mouse liver. Toxicol Appl Pharmacol 113(2):293–298

Min KS, Kim H, Fujii M, Tetsuchikawahara N, Onosaka S (2002) Glucocorticoids suppress the inflammation-mediated tolerance to acute toxicity of cadmium in mice. Toxicol Appl Pharmacol 178(1):1–7

Misra RR, Hochadel JF, Smith GT, Cook JC, Waalkes MP, Wink DA (1996) Evidence that nitric oxide enhances cadmium toxicity by displacing the metal from metallothionein. Chem Res Toxicol 9(1):326–332

Mita M, Imura N, Kumazawa Y, Himeno S (2002) Suppressed proliferative response of spleen T cells from metallothionein null mice. Microbiol Immunol 46(2):101–107

Mita M, Satoh M, Shimada A, Okajima M, Azuma S, Suzuki JS, Sakabe K, Hara S, Himeno S (2008) Metallothionein is a crucial protective factor against Helicobacter pylori-induced gastric erosive lesions in a mouse model. Am J Physiol Gastrointest Liver Physiol 294(4): G877–884

Mita M, Satoh M, Shimada A, Azuma S, Himeno S, Hara S (2012) Metallothionein deficiency exacerbates chronic inflammation associated with carcinogenesis in stomach of mice infected with Helicobacter pylori. J Toxicol Sci 37(6):1261–1265

Mitani T, Shirasaka D, Aoyama N, Miki I, Morita Y, Ikehara N, Matsumoto Y, Okuno T, Toyoda M, Miyachi H, Yoshida S, Chayahara N, Hori J, Tamura T, Azuma T, Kasuga M (2008) Role of metallothionein in Helicobacter pylori-positive gastric mucosa with or without early gastric cancer and the effect on its expression after eradication therapy. J Gastroenterol Hepatol 23(8 Pt 2):e334–339

Mitropoulos D, Kyroudi-Voulgari A, Theocharis S, Serafetinides E, Moraitis E, Zervas A, Kittas C (2005) Prognostic significance of metallothionein expression in renal cell carcinoma. World J Surg Oncol 3(1):5

Moffatt P, Séguin C (1998) Expression of the gene encoding metallothionein-3 in organs of the reproductive system. DNA Cell Biol 17(6):501–510

Moltedo O, Verde C, Capasso A, Parisi E, Remondelli P, Bonatti S, Alvarez-Hernandez X, Glass J, Alvino CG, Leone A (2000) Zinc transport and metallothionein secretion in the intestinal human cell line Caco-2. J Biol Chem 275(41):31819–31825

Monnet-Tschudi F, Zurich MG, Boschat C, Corbaz A, Honegger P (2006) Involvement of environmental mercury and lead in the etiology of neurodegenerative diseases. Rev Environ Health 21(2):105–117

Morandi L, de Biase D, Visani M, Monzoni A, Tosi A, Brulatti M, Turchetti D, Baccarini P, Tallini G, Pession A (2012) T([20]) repeat in the 3'-untranslated region of the MT1X gene: a marker with high sensitivity and specificity to detect microsatellite instability in colorectal cancer. Int J Colorectal Dis 27(5):647–656

Morellini NM, Giles NL, Rea S, Adcroft KF, Falder S, King CE, Dunlop SA, Beazley LD, West AK, Wood FM, Fear MW (2008) Exogenous metallothionein-IIA promotes accelerated healing after a burn wound. Wound Repair Regen 16(5):682–690

Morinobu A, Kanno Y, O'Shea JJ (2004) Discrete roles for histone acetylation in human T helper 1 cell-specific gene expression. J Biol Chem 279(39):40640–40646

Mounicou S, Ouerdane L, L'Azou B, Passagne I, Ohayon-Courtes C, Szpunar J, Lobinski R (2010) Identification of metallothionein subisoforms in HPLC using accurate mass and online sequencing by electrospray hybrid linear ion trap-orbital ion trap mass spectrometry. Anal Chem 82(16):6947–6957

Moussa M, Kloth D, Peers G, Cherian MG, Frei JV, Chin JL (1997) Metallothionein expression in prostatic carcinoma: correlation with Gleason grade, pathologic stage, DNA content and serum level of prostate-specific antigen. Clin Invest Med 20(6):371–380

Mukhopadhyay D, Mitra A, Nandi P, Varghese AC, Murmu N, Chowdhury R, Chaudhuri K, Bhattacharyya AK (2009) Expression of metallothionein-1 (MT-1) mRNA in the rat testes and liver after cadmium injection. Syst Biol Reprod Med 55(5–6):188–192

Muramatsu Y, Hasegawa Y, Fukano H, Ogawa T, Namuba M, Mouri K, Fujimoto Y, Matsuura H, Takai Y, Mori M (2000) Metallothionein immunoreactivity in head and neck carcinomas; special reference to clinical behaviors and chemotherapy responses. Anticancer Res 20 (1A):257–264

Murata M, Gong P, Suzuki K, Koizumi S (1999) Differential metal response and regulation of human heavy metal-inducible genes. J Cell Physiol 180(1):105–113

Murphy BJ, Kimura T, Sato BG, Shi Y, Andrews GK (2008) Metallothionein induction by hypoxia involves cooperative interactions between metal-responsive transcription factor-1 and hypoxia-inducible transcription factor-1alpha. Mol Cancer Res 6(3):483–490

Nagano T, Itoh N, Ebisutani C, Takatani T, Miyoshi T, Nakanishi T, Tanaka K (2000) The transport mechanism of metallothionein is different from that of classical NLS-bearing protein. J Cell Physiol 185(3):440–446

Nagata T, Takahashi Y, Ishii Y, Asai S, Nishida Y, Murata A, Koshinaga T, Fukuzawa M, Hamazaki M, Asami K, Ito E, Ikeda H, Takamatsu H, Koike K, Kikuta A, Kuroiwa M, Watanabe A, Kosaka Y, Fujita H, Miyake M, Mugishima H (2003) Transcriptional profiling in hepatoblastomas using high-density oligonucleotide DNA array. Cancer Genet Cytogenet 145(2):152–160

Nagel WW, Vallee BL (1995) Cell cycle regulation of metallothionein in human colonic cancer cells. Proc Natl Acad Sci USA 92(2):579–583

Nakajima K, Suzuki K (1995) Immunochemical detection of metallothionein in brain. Neurochem Int 27(1):73–87

Nakano M, Sogawa CA, Sogawa N, Mishima K, Yamachika E, Mizukawa N, Fukunaga J, Kawamoto T, Sawaki K, Sugahara T, Furuta H (2003) Expression pattern of cisplatin-induced metallothionein isoforms in squamous cell carcinoma. Anticancer Res 23(1A):299–303

Nakayama A, Fukuda H, Ebara M, Hamasaki H, Nakajima K, Sakurai H (2002) A new diagnostic method for chronic hepatitis, liver cirrhosis, and hepatocellular carcinoma based on serum metallothionein, copper, and zinc levels. Biol Pharm Bull 25(4):426–431

Nartey N, Cherian MG, Banerjee D (1987a) Immunohistochemical localization of metallothionein in human thyroid tumors. Am J Pathol 129(1):177–182

Nartey NO, Frei JV, Cherian MG (1987b) Hepatic copper and metallothionein distribution in Wilson's disease (hepatolenticular degeneration). Lab Invest 57(4):397–401

Naruse S, Igarashi S, Furuya T, Kobayashi H, Miyatake T, Tsuji S (1994) Structures of the human and mouse growth inhibitory factor-encoding genes. Gene 144(2):283–287

Nemoto K, Kondo Y, Himeno S, Suzuki Y, Hara S, Akimoto M, Imura N (2000) Modulation of telomerase activity by zinc in human prostatic and renal cancer cells. Biochem Pharmacol 59 (4):401–405

Ngu TT, Stillman MJ (2006) Arsenic binding to human metallothionein. J Am Chem Soc 128 (38):12473–12483

Ngu TT, Stillman MJ (2009) Metalation of metallothioneins. IUBMB Life 61(4):438–446

Ngu TT, Dryden MD, Stillman MJ (2010a) Arsenic transfer between metallothionein proteins at physiological pH. Biochem Biophys Res Commun 401(1):69–74

Ngu TT, Krecisz S, Stillman MJ (2010b) Bismuth binding studies to the human metallothionein using electrospray mass spectrometry. Biochem Biophys Res Commun 396(2):206–212

Nguyen A, Jing Z, Mahoney PS, Davis R, Sikka SC, Agrawal KC, Abdel-Mageed AB (2000) In vivo gene expression profile analysis of metallothionein in renal cell carcinoma. Cancer Lett 160(2):133–140

Nguyen T, Sherratt PJ, Pickett CB (2003) Regulatory mechanisms controlling gene expression mediated by the antioxidant response element. Annu Rev Pharmacol Toxicol 43:233–260

Ni FY, Cai B, Ding ZC, Zheng F, Zhang MJ, Wu HM, Sun HZ, Huang ZX (2007) Structural prediction of the beta-domain of metallothionein-3 by molecular dynamics simulation. Proteins 68(1):255–266

Nielsen AE, Bohr A, Penkowa M (2007) The Balance between Life and Death of Cells: Roles of Metallothioneins. Biomark Insights 7(1):99–111

Nielson KB, Atkin CL, Winge DR (1985) Distinct metal-binding configurations in metallothionein. J Biol Chem 260(9):5342–5350

Nishimura N, Reeve VE, Nishimura H, Satoh M, Tohyama C (2000) Cutaneous metallothionein induction by ultraviolet B irradiation in interleukin-6 null mice. J Invest Dermatol 114 (2):343–348

Nomiya T, Nemoto K, Nakata E, Miyachi H, Takai Y, Yamada S (2004) Intrinsic radiosensitivity by metallothionein expression has no great influence on clinical radiosensitivity in esophageal carcinoma. Oncol Rep 12(6):1195–1199

Nordberg M (1998) Metallothioneins: historical review and state of knowledge. Talanta 46 (2):243–254

Nordberg M, Nordberg GF (2000) Toxicological aspects of metallothionein. Cell Mol Biol (Noisy-le-grand) 46(2):451–463

Nordberg GF, Goyer R, Nordberg M (1975) Comparative toxicity of cadmium-metallothionein and cadmium chloride on mouse kidney. Arch Pathol 99(4):192–197

Nordberg GF, Garvey JS, Chang CC (1982) Metallothionein in plasma and urine of cadmium workers. Environ Res 28(1):179–182

Nzengue Y, Lefebvre E, Cadet J, Favier A, Rachidi W, Steiman R, Guiraud P (2009) Metallothionein expression in HaCaT and C6 cell lines exposed to cadmium. J Trace Elem Med Biol 23(4):314–323

Oberbarnscheidt J, Kind P, Abel J, Gleichmann E (1988) Metallothionein induction in a human B cell line by stimulated immune cell products. Res Commun Chem Pathol Pharmacol 60 (2):211–224

O'Connor RA, Prendergast CT, Sabatos CA, Lau CW, Leech MD, Wraith DC, Anderton SM (2008) Cutting edge: Th1 cells facilitate the entry of Th17 cells to the central nervous system during experimental autoimmune encephalomyelitis. J Immunol 181(6):3750–3754

Oda N, Sogawa CA, Sogawa N, Onodera K, Furuta H, Yamamoto T (2001) Metallothionein expression and localization in rat bone tissue after cadmium injection. Toxicol Lett 123 (2–3):143–150

Ogawa Y (2003) Immunocytochemistry of myoepithelial cells in the salivary glands. Prog Histochem Cytochem 38(4):343–426

Ogra Y, Suzuki KT (2000) Nuclear trafficking of metallothionein: possible mechanisms and current knowledge. Cell Mol Biol (Noisy-le-grand) 46(2):357–365

Ohshio G, Imamura T, Okada N, Wang ZH, Yamaki K, Kyogoku T, Suwa H, Yamabe H, Imamura M (1996) Immunohistochemical study of metallothionein in pancreatic carcinomas. J Cancer Res Clin Oncol 122(6):351–355

Oka D, Yamashita S, Tomioka T, Nakanishi Y, Kato H, Kaminishi M, Ushijima T (2009) The presence of aberrant DNA methylation in noncancerous esophageal mucosae in association with smoking history: a target for risk diagnosis and prevention of esophageal cancers. Cancer 115(15):3412–3426

Olesen C, Moller M, Byskov AG (2004) Tesmin transcription is regulated differently during male and female meiosis. Mol Reprod Dev 67(1):116–126

Ono S, Hirai K, Tokuda E (2009) Effects of pergolide mesilate on metallothionein mRNAs expression in a mouse model for Parkinson disease. Biol Pharm Bull 32(10):1813–1817

Ostrakhovitch EA, Cherian MG (2004) Differential regulation of signal transduction pathways in wild type and mutated p53 breast cancer epithelial cells by copper and zinc. Arch Biochem Biophys 423(2):351–361

Ostrakhovitch EA, Olsson PE, Jiang S, Cherian MG (2006) Interaction of metallothionein with tumor suppressor p53 protein. FEBS Lett 580(5):1235–1238

Ostrakhovitch EA, Olsson PE, von Hofsten J, Cherian MG (2007) P53 mediated regulation of metallothionein transcription in breast cancer cells. J Cell Biochem 102(6):1571–1583

Otsuka F, Okugaito I, Ohsawa M, Iwamatsu A, Suzuki K, Koizumi S (2000) Novel responses of ZRF, a variant of human MTF-1, to in vivo treatment with heavy metals. Biochim Biophys Acta 1492(2–3):330–340

Otsuka T, Hamada A, Iguchi K, Usui S, Hirano K (2013) Suppression of metallothionein 3 gene expression by androgen in LNCaP prostate cancer cells. Biomed Rep 1(4):614–618

Oyama T, Take H, Hikino T, Iino Y, Nakajima T (1996) Immunohistochemical expression of metallothionein in invasive breast cancer in relation to proliferative activity, histology and prognosis. Oncology 53(2):112–117

Oz G, Zangger K, Armitage IM (2001) Three-dimensional structure and dynamics of a brain specific growth inhibitory factor: metallothionein-3. Biochemistry 40(38):11433–11441

Ozer H, Yenicesu G, Arici S, Cetin M, Tuncer E, Cetin A (2012) Immunohistochemistry with apoptotic-antiapoptotic proteins (p53, p21, bax, bcl-2), c-kit, telomerase, and metallothionein as a diagnostic aid in benign, borderline, and malignant serous and mucinous ovarian tumors. Diagn Pathol 7:124

Palacios O, Atrian S, Capdevila M (2011) Zn- and Cu-thioneins: a functional classification for metallothioneins? J Biol Inorg Chem 16(7):991–1009

Palecek E, Brazdova M, Cernocka H, Vlk D, Brazda V, Vojtesek B (1999) Effect of transition metals on binding of p53 protein to supercoiled DNA and to consensus sequence in DNA fragments. Oncogene 18(24):3617–3625

Palmiter RD (1995) Constitutive expression of metallothionein-III (MT-III), but not MT-I, inhibits growth when cells become zinc deficient. Toxicol Appl Pharmacol 135(1):139–146

Palmiter RD (1998) The elusive function of metallothioneins. Proc Natl Acad Sci USA 95 (15):8428–8430

Palmiter RD, Findley SD, Whitmore TE, Durnam DM (1992) MT-III, a brain-specific member of the metallothionein gene family. Proc Natl Acad Sci USA 89(14):6333–6337

Palumaa P, Njunkova O, Pokras L, Eriste E, Jornvall H, Sillard R (2002) Evidence for non-isostructural replacement of Zn(2+) with Cd(2+) in the beta-domain of brain-specific metallothionein-3. FEBS Lett 527(1–3):76–80

Palumaa P, Eriste E, Kruusel K, Kangur L, Jornvall H, Sillard R (2003) Metal binding to brain-specific metallothionein-3 studied by electrospray ionization mass spectrometry. Cell Mol Biol (Noisy-le-grand) 49(5):763–768

Palumaa P, Tammiste I, Kruusel K, Kangur L, Jornvall H, Sillard R (2005) Metal binding of metallothionein-3 versus metallothionein-2: lower affinity and higher plasticity. Biochim Biophys Acta 1747(2):205–211

Pan Y, Huang J, Xing R, Yin X, Cui J, Li W, Yu J, Lu Y (2013a) Metallothionein 2A inhibits NF-kappaB pathway activation and predicts clinical outcome segregated with TNM stage in gastric cancer patients following radical resection. J Transl Med 11:173

Pan YM, Xing R, Cui JT, Li WM, Lu YY (2013b) Clinicopathological significance of altered metallothionein 2A expression in gastric cancer according to Lauren's classification. Chin Med J (Engl) 126(14):2681–2686

Panani AD (2008) Cytogenetic and molecular aspects of gastric cancer: clinical implications. Cancer Lett 266(2):99–115

Pankhurst MW, Bennett W, Kirkcaldie MT, West AK, Chung RS (2011) Increased circulating leukocyte numbers and altered macrophage phenotype correlate with the altered immune response to brain injury in metallothionein (MT)-I/II null mutant mice. J Neuroinflammation 8:172

Papouli E, Defais M, Larminat F (2002) Overexpression of metallothionein-II sensitizes rodent cells to apoptosis induced by DNA cross-linking agent through inhibition of NF-kappa B activation. J Biol Chem 277(7):4764–4769

Park SY, Kwon HJ, Lee HE, Ryu HS, Kim SW, Kim JH, Kim IA, Jung N, Cho NY, Kang GH (2011) Promoter CpG island hypermethylation during breast cancer progression. Virchows Arch 458(1):73–84

Park YH, Lee YM, Kim DS, Park J, Suk K, Kim JK, Han HS (2013) Hypothermia enhances induction of protective protein metallothionein under ischemia. J Neuroinflammation 10:21

Pastuszewski W, Dziegiel P, Krecicki T, Podhorska-Okolow M, Ciesielska U, Gorzynska E, Zabel M (2007) Prognostic significance of metallothionein, p53 protein and Ki-67 antigen expression in laryngeal cancer. Anticancer Res 27(1A):335–342

Pearson CA, Lamar PC, Prozialeck WC (2003) Effects of cadmium on E-cadherin and VE-cadherin in mouse lung. Life Sci 72(11):1303–1320

Pedersen MO, Larsen A, Stoltenberg M, Penkowa M (2009) The role of metallothionein in oncogenesis and cancer prognosis. Prog Histochem Cytochem 44(1):29–64

Peixoto NC, Serafim MA, Flores EM, Bebianno MJ, Pereira ME (2007) Metallothionein, zinc, and mercury levels in tissues of young rats exposed to zinc and subsequently to mercury. Life Sci 81(16):1264–1271

Pena OM, Pistolic J, Raj D, Fjell CD, Hancock RE (2011) Endotoxin tolerance represents a distinctive state of alternative polarization (M2) in human mononuclear cells. J Immunol 186 (12):7243–7254

Peng D, Hu TL, Jiang A, Washington MK, Moskaluk CA, Schneider-Stock R, El-Rifai W (2011) Location-specific epigenetic regulation of the metallothionein 3 gene in esophageal adenocarcinomas. PLoS One 6(7), e22009

Penkowa M, Hidalgo J (2000) Metallothionein I+II expression and their role in experimental autoimmune encephalomyelitis. Glia 32(3):247–263

Penkowa M, Hidalgo J (2001) Metallothionein treatment reduces proinflammatory cytokines IL-6 and TNF-alpha and apoptotic cell death during experimental autoimmune encephalomyelitis (EAE). Exp Neurol 170(1):1–14

Penkowa M, Giralt M, Moos T, Thomsen PS, Hernandez J, Hidalgo J (1999a) Impaired inflammatory response to glial cell death in genetically metallothionein-I- and -II-deficient mice. Exp Neurol 156(1):149–164

Penkowa M, Moos T, Carrasco J, Hadberg H, Molinero A, Bluethmann H, Hidalgo J (1999b) Strongly compromised inflammatory response to brain injury in interleukin-6-deficient mice. Glia 25(4):343–357

Penkowa M, Espejo C, Martinez-Caceres EM, Poulsen CB, Montalban X, Hidalgo J (2001) Altered inflammatory response and increased neurodegeneration in metallothionein I+II deficient mice during experimental autoimmune encephalomyelitis. J Neuroimmunol 119 (2):248–260

Penkowa M, Espejo C, Ortega-Aznar A, Hidalgo J, Montalban X, Martinez Caceres EM (2003) Metallothionein expression in the central nervous system of multiple sclerosis patients. Cell Mol Life Sci 60(6):1258–1266

Penkowa M, Tio L, Giralt M, Quintana A, Molinero A, Atrian S, Vasak M, Hidalgo J (2006) Specificity and divergence in the neurobiologic effects of different metallothioneins after brain injury. J Neurosci Res 83(6):974–984

Penkowa M, Sørensen BL, Nielsen SL, Hansen PB (2009) Metallothionein as a useful marker in Hodgkin lymphoma subclassification. Leuk Lymphoma 50(2):200–210

Perego P, Romanelli S, Carenini N, Magnani I, Leone R, Bonetti A, Paolicchi A, Zunino F (1998) Ovarian cancer cisplatin-resistant cell lines: multiple changes including collateral sensitivity to Taxol. Ann Oncol 9(4):423–430

Perez-Gutierrez S, Gonzalez-Campora R, Amerigo-Navarro J, Beato-Moreno A, Sanchez-Leon M, Pareja Megia JM, Virizuela-Echaburu JA, Lopez-Beltran A (2007) Expression of P-glycoprotein and metallothionein in gastrointestinal stromal tumor and leiomyosarcomas. Clinical implications. Pathol Oncol Res 13(3):203–208

Perkins ND (2007) Integrating cell-signalling pathways with NF-kappaB and IKK function. Nat Rev Mol Cell Biol 8(1):49–62

Person RJ, Tokar EJ, Xu Y, Orihuela R, Ngalame NN, Waalkes MP (2013) Chronic cadmium exposure in vitro induces cancer cell characteristics in human lung cells. Toxicol Appl Pharmacol 273(2):281–288

Petering DH, Huang M, Moteki S, Shaw CF 3rd (2000) Cadmium and lead interactions with transcription factor IIIA from Xenopus laevis: a model for zinc finger protein reactions with toxic metal ions and metallothionein. Mar Environ Res 50(1–5):89–92

Philcox JC, Coyle P, Michalska A, Choo KH, Rofe AM (1995) Endotoxin-induced inflammation does not cause hepatic zinc accumulation in mice lacking metallothionein gene expression. Biochem J 308(Pt 2):543–546

Podhorska-Okolow M, Dziegiel P, Dolinska-Krajewska B, Dumanska M, Cegielski M, Jethon Z, Rossini K, Carraro U, Zabel M (2006) Expression of metallothionein in renal tubules of rats exposed to acute and endurance exercise. Folia Histochem Cytobiol 44(3):195–200

Pontes HA, de Aquino Xavier FC, da Silva TS, Fonseca FP, Paiva HB, Pontes FS, dos Santos PD Jr (2009) Metallothionein and p-Akt proteins in oral dysplasia and in oral squamous cell carcinoma: an immunohistochemical study. J Oral Pathol Med 38(8):644–650

Potter EG, Cheng Y, Knight JB, Gordish-Dressman H, Natale JE (2007) Basic science; metallothionein I and II attenuate the thalamic microglial response following traumatic axotomy in the immature brain. J Neurotrauma 24(1):28–42

Potts RJ, Bespalov IA, Wallace SS, Melamede RJ, Hart BA (2001) Inhibition of oxidative DNA repair in cadmium-adapted alveolar epithelial cells and the potential involvement of metallothionein. Toxicology 161(1–2):25–38

Poulsen CB, Borup R, Borregaard N, Nielsen FC, Møller MB, Ralfkiaer E (2006) Prognostic significance of metallothionein in B-cell lymphomas. Blood 108(10):3514–3519

Powell SR (2000) The antioxidant properties of zinc. J Nutr 130(5S Suppl):1447S–1454S

Pula B, Domoslawski P, Podhorska-Okolow M, Dziegiel P (2012) Role of metallothioneins in benign and malignant thyroid lesions. Thyroid Res 5(1):26

Pula B, Strutynska-Karpinska M, Markowska-Woyciechowska A, Jethon A, Wolowiec D, Rys J, Podhorska-Okolow M, Grabowski K, Dziegiel P (2013) Expression of metallothionein and Ki-67 antigen in GISTs of different grade of malignancy. Adv Clin Exp Med 22(4):513–518

Pula B, Tazbierski T, Zamirska A, Werynska B, Bieniek A, Szepietowski J, Rys J, Dziegiel P, Podhorska-Okolow M (2015) Metallothionein 3 Expression in Normal Skin and Malignant Skin Lesions. Pathol Oncol Res 21:187–193

Qu W, Diwan BA, Liu J, Goyer RA, Dawson T, Horton JL, Cherian MG, Waalkes MP (2002) The metallothionein-null phenotype is associated with heightened sensitivity to lead toxicity and an inability to form inclusion bodies. Am J Pathol 160(3):1047–1056

Quaife CJ, Findley SD, Erickson JC, Froelick GJ, Kelly EJ, Zambrowicz BP, Palmiter RD (1994) Induction of a new metallothionein isoform (MT-IV) occurs during differentiation of stratified squamous epithelia. Biochemistry 33(23):7250–7259

Quaife CJ, Cherne RL, Newcomb TG, Kapur RP, Palmiter RD (1999) Metallothionein overexpression suppresses hepatic hyperplasia induced by hepatitis B surface antigen. Toxicol Appl Pharmacol 155(2):107–116

Quesada AR, Byrnes RW, Krezoski SO, Petering DH (1996) Direct reaction of H2O2 with sulfhydryl groups in HL-60 cells: zinc-metallothionein and other sites. Arch Biochem Biophys 334(2):241–250

Rae MT, Niven D, Ross A, Forster T, Lathe R, Critchley HO, Ghazal P, Hillier SG (2004) Steroid signalling in human ovarian surface epithelial cells: the response to interleukin-1alpha determined by microarray analysis. J Endocrinol 183(1):19–28

Raes G, De Baetselier P, Noel W, Beschin A, Brombacher F, Hassanzadeh Gh G (2002) Differential expression of FIZZ1 and Ym1 in alternatively versus classically activated macrophages. J Leukoc Biol 71(4):597–602

Rahman I, Marwick J, Kirkham P (2004) Redox modulation of chromatin remodeling: impact on histone acetylation and deacetylation, NF-kappaB and pro-inflammatory gene expression. Biochem Pharmacol 68(6):1255–1267

Raleigh JA, Chou SC, Calkins-Adams DP, Ballenger CA, Novotny DB, Varia MA (2000) A clinical study of hypoxia and metallothionein protein expression in squamous cell carcinomas. Clin Cancer Res 6(3):855–862

Rana U, Kothinti R, Meeusen J, Tabatabai NM, Krezoski S, Petering DH (2008) Zinc binding ligands and cellular zinc trafficking: apo-metallothionein, glutathione, TPEN, proteomic zinc, and Zn-Sp1. J Inorg Biochem 102(3):489–499

Rani A, Kumar A, Lal A, Pant M (2014) Cellular mechanisms of cadmium-induced toxicity: a review. Int J Environ Health Res 24(4):378–399

Raudenska M, Gumulec J, Podlaha O, Sztalmachova M, Babula P, Eckschlager T, Adam V, Kizek R, Masarik M (2014) Metallothionein polymorphisms in pathological processes. Metallomics 6(1):55–68

Raymond AD, Gekonge B, Giri MS, Hancock A, Papasavvas E, Chehimi J, Kossenkov AV, Nicols C, Yousef M, Mounzer K, Shull J, Kostman J, Showe L, Montaner LJ (2010) Increased metallothionein gene expression, zinc, and zinc-dependent resistance to apoptosis in circulating monocytes during HIV viremia. J Leukoc Biol 88(3):589–596

Reeve VE, Nishimura N, Bosnic M, Michalska AE, Choo KH (2000) Lack of metallothionein-I and -II exacerbates the immunosuppressive effect of ultraviolet B radiation and cis-urocanic acid in mice. Immunology 100(3):399–404

Regunathan A, Glesne DA, Wilson AK, Song J, Nicolae D, Flores T, Bhattacharyya MH (2003) Microarray analysis of changes in bone cell gene expression early after cadmium gavage in mice. Toxicol Appl Pharmacol 191(3):272–293

Reinecke F, Levanets O, Olivier Y, Louw R, Semete B, Grobler A, Hidalgo J, Smeitink J, Olckers A, Van der Westhuizen FH (2006) Metallothionein isoform 2A expression is inducible and protects against ROS-mediated cell death in rotenone-treated HeLa cells. Biochem J 395 (2):405–415

Rekers NV, Bajema IM, Mallat MJ, Anholts JD, de Vaal YJ, Zandbergen M, Haasnoot GW, van Zwet EW, de Fijter JW, Claas FH, Eikmans M (2013) Increased metallothionein expression reflects steroid resistance in renal allograft recipients. Am J Transplant 13(8):2106–2118

Rizkalla KS, Cherian MG (1997) Metallothionein: a potential marker for differentiating benign and neoplastic gastrointestinal lymphoid infiltrates. Pathology 29(2):141–146

Roesijadi G, Bogumil R, Vasak M, Kagi JH (1998) Modulation of DNA binding of a tramtrack zinc finger peptide by the metallothionein-thionein conjugate pair. J Biol Chem 273 (28):17425–17432

Rofe AM, Philcox JC, Coyle P (1996) Trace metal, acute phase and metabolic response to endotoxin in metallothionein-null mice. Biochem J 314(Pt 3):793–797

Romero-Isart N, Vasak M (2002) Advances in the structure and chemistry of metallothioneins. J Inorg Biochem 88(3–4):388–396

Roth MJ, Abnet CC, Hu N, Wang QH, Wei WQ, Green L, D'Alelio M, Qiao YL, Dawsey SM, Taylor PR, Woodson K (2006) p16, MGMT, RARbeta2, CLDN3, CRBP and MT1G gene methylation in esophageal squamous cell carcinoma and its precursor lesions. Oncol Rep 15 (6):1591–1597

Rudolf E, Cervinka M (2010) Zinc pyrithione induces cellular stress signaling and apoptosis in Hep-2 cervical tumor cells: the role of mitochondria and lysosomes. Biometals 23(2):339–354

Ruiz-Riol M, Martinez-Arconada MJ, Alonso N, Soldevila B, Marchena D, Armengol MP, Sanmarti A, Pujol-Borrell R, Martinez-Caceres EM (2012) Overexpression of metallothionein I/II: a new feature of thyroid follicular cells in Graves' disease. J Clin Endocrinol Metab 97 (2):446–454

Ryu HH, Jung S, Jung TY, Moon KS, Kim IY, Jeong YI, Jin SG, Pei J, Wen M, Jang WY (2012) Role of metallothionein 1E in the migration and invasion of human glioma cell lines. Int J Oncol 41(4):1305–1313

Sabolic I (2006) Common mechanisms in nephropathy induced by toxic metals. Nephron Physiol 104(3):p107–114

Sabolic I, Breljak D, Skarica M, Herak-Kramberger CM (2010) Role of metallothionein in cadmium traffic and toxicity in kidneys and other mammalian organs. Biometals 23 (5):897–926

Safieh-Garabedian B, Mayasi Y, Saade NE (2012) Targeting neuroinflammation for therapeutic intervention in neurodegenerative pathologies: a role for the peptide analogue of thymulin (PAT). Expert Opin Ther Targets 16(11):1065–1073

Saga Y, Hashimoto H, Yachiku S, Tokumitsu M, Kaneko S (2002) Immunohistochemical expression of metallothionein in human bladder cancer: correlation with histopathological parameters and patient survival. J Urol 168(5):2227–2231

Saga Y, Hashimoto H, Yachiku S, Iwata T, Tokumitsu M (2004) Reversal of acquired cisplatin resistance by modulation of metallothionein in transplanted murine tumors. Int J Urol 11 (6):407–415

Saika T, Tsushima T, Nasu Y, Akebi N, Noda M, Kobashi K, Matsumura Y, Ohmori H (1992) Histopathological study of metallothionein in bladder cancer and renal cell carcinoma. Nihon Hinyokika Gakkai Zasshi 83(5):636–642

Saika T, Tsushima T, Ochi J, Akebi N, Nasu Y, Matsumura Y, Ohmori H (1994) Over-expression of metallothionein and drug-resistance in bladder cancer. Int J Urol 1(2):135–139

Saito C, Yan HM, Artigues A, Villar MT, Farhood A, Jaeschke H (2010) Mechanism of protection by metallothionein against acetaminophen hepatotoxicity. Toxicol Appl Pharmacol 242 (2):182–190

Sakamoto LH, de Camargo B, Cajaiba M, Soares FA, Vettore AL (2010) MT1G hypermethylation: a potential prognostic marker for hepatoblastoma. Pediatr Res 67 (4):387–393

Sakurai A, Hara S, Okano N, Kondo Y, Inoue J, Imura N (1999) Regulatory role of metallothionein in NF-kappaB activation. FEBS Lett 455(1–2):55–58

Santon A, Formigari A, Albergoni V, Irato P (2006) Effect of Zn treatment on wild type and MT-null cell lines in relation to apoptotic and/or necrotic processes and on MT isoform gene expression. Biochim Biophys Acta 1763(3):305–312

Sato M, Bremner I (1993) Oxygen free radicals and metallothionein. Free Radic Biol Med 14 (3):325–337

Sato M, Kondoh M (2002) Recent studies on metallothionein: protection against toxicity of heavy metals and oxygen free radicals. Tohoku J Exp Med 196(1):9–22

Sato M, Sasaki M, Hojo H (1994) Differential induction of metallothionein synthesis by interleukin-6 and tumor necrosis factor-alpha in rat tissues. Int J Immunopharmacol 16 (2):187–195

Satoh M, Cherian MG, Imura N, Shimizu H (1994) Modulation of resistance to anticancer drugs by inhibition of metallothionein synthesis. Cancer Res 54(20):5255–5257

Sauer JM, Waalkes MP, Hooser SB, Baines AT, Kuester RK, Sipes IG (1997) Tolerance induced by all-trans-retinol to the hepatotoxic effects of cadmium in rats: role of metallothionein expression. Toxicol Appl Pharmacol 143(1):110–119

Sauerbrey A, Zintl F, Hermann J, Volm M (1998) Multiple resistance mechanisms in acute nonlymphoblastic leukemia (ANLL). Anticancer Res 18(2B):1231–1236

Savino W, Huang PC, Corrigan A, Berrih S, Dardenne M (1984) Thymic hormone-containing cells. V. Immunohistological detection of metallothionein within the cells bearing thymulin (a zinc-containing hormone) in human and mouse thymuses. J Histochem Cytochem 32 (9):942–946

Saydam N, Adams TK, Steiner F, Schaffner W, Freedman JH (2002) Regulation of metallothionein transcription by the metal-responsive transcription factor MTF-1: identification of signal transduction cascades that control metal-inducible transcription. J Biol Chem 277 (23):20438–20445

Schlake T, Boehm T (2001) Expression domains in the skin of genes affected by the nude mutation and identified by gene expression profiling. Mech Dev 109(2):419–422

Schmitz KJ, Lang H, Kaiser G, Wohlschlaeger J, Sotiropoulos GC, Baba HA, Jasani B, Schmid KW (2009) Metallothionein overexpression and its prognostic relevance in intrahepatic cholangiocarcinoma and extrahepatic hilar cholangiocarcinoma (Klatskin tumors). Hum Pathol 40(12):1706–1714

Schuhmacher-Wolz U, Dieter HH, Klein D, Schneider K (2009) Oral exposure to inorganic arsenic: evaluation of its carcinogenic and non-carcinogenic effects. Crit Rev Toxicol 39 (4):271–298

Schwarz MA, Lazo JS, Yalowich JC, Allen WP, Whitmore M, Bergonia HA, Tzeng E, Billiar TR, Robbins PD, Lancaster JR Jr et al (1995) Metallothionein protects against the cytotoxic and DNA-damaging effects of nitric oxide. Proc Natl Acad Sci USA 92(10):4452–4456

Seibold P, Hein R, Schmezer P, Hall P, Liu J, Dahmen N, Flesch-Janys D, Popanda O, Chang-Claude J (2011) Polymorphisms in oxidative stress-related genes and postmenopausal breast cancer risk. Int J Cancer 129(6):1467–1476

Seibold P, Hall P, Schoof N, Nevanlinna H, Heikkinen T, Benner A, Liu J, Schmezer P, Popanda O, Flesch-Janys D, Chang-Claude J (2013) Polymorphisms in oxidative stress-related genes and mortality in breast cancer patients--potential differential effects by radiotherapy? Breast 22(5):817–823

Sens MA, Somji S, Lamm DL, Garrett SH, Slovinsky F, Todd JH, Sens DA (2000) Metallothionein isoform 3 as a potential biomarker for human bladder cancer. Environ Health Perspect 108(5):413–418

Sens MA, Somji S, Garrett SH, Beall CL, Sens DA (2001) Metallothionein isoform 3 overexpression is associated with breast cancers having a poor prognosis. Am J Pathol 159 (1):21–26

Sewell AK, Jensen LT, Erickson JC, Palmiter RD, Winge DR (1995) Bioactivity of metallothionein-3 correlates with its novel beta domain sequence rather than metal binding properties. Biochemistry 34(14):4740–4747

Sgagias M, Balter NJ, Gray I (1989) Uptake and subcellular distribution of cadmium in resting and mitogen-activated lymphocytes and its relationship to a metallothionein-like protein. Environ Res 49(2):262–270

Sharma S, Ebadi M (2014) Significance of metallothioneins in aging brain. Neurochem Int 65:40–48

Shimoda R, Achanzar WE, Qu W, Nagamine T, Takagi H, Mori M, Waalkes MP (2003) Metallothionein is a potential negative regulator of apoptosis. Toxicol Sci 73(2):294–300

Shiraishi N, Rehm S, Waalkes MP (1994) Effect of chlorpromazine pretreatment on cadmium toxicity in the male Wistar (WF/NCr) rat. J Toxicol Environ Health 42(2):193–208

Shnyder SD, Hayes AJ, Pringle J, Archer CW (1998) P-glycoprotein and metallothionein expression and resistance to chemotherapy in osteosarcoma. Br J Cancer 78(6):757–759

Siegsmund MJ, Marx C, Seemann O, Schummer B, Steidler A, Toktomambetova L, Köhrmann KU, Rassweiler J, Alken P (1999) Cisplatin-resistant bladder carcinoma cells: enhanced expression of metallothioneins. Urol Res 27(3):157–163

Simpkins C, Lloyd T, Li S, Balderman S (1998) Metallothionein-induced increase in mitochondrial inner membrane permeability. J Surg Res 75(1):30–34

Sliwinska-Mosson M, Milnerowicz H, Rabczynski J, Milnerowicz S (2009) Immunohistochemical localization of metallothionein and p53 protein in pancreatic serous cystadenomas. Arch Immunol Ther Exp (Warsz) 57(4):295–301

Sliwinska-Mosson M, Milnerowicz H, Jablonowska M, Milnerowicz S, Nabzdyk S, Rabczynski J (2012) The effect of smoking on expression of IL-6 and antioxidants in pancreatic fluids and tissues in patients with chronic pancreatitis. Pancreatology 12(4):295–304

Smaoui-Damak W, Berthet B, Hamza-Chaffai A (2009) In situ potential use of metallothionein as a biomarker of cadmium contamination in Ruditapes decussatus. Ecotoxicol Environ Saf 72 (5):1489–1498

Smirnova IV, Bittel DC, Ravindra R, Jiang H, Andrews GK (2000) Zinc and cadmium can promote rapid nuclear translocation of metal response element-binding transcription factor-1. J Biol Chem 275(13):9377–9384

Smith E, Drew PA, Tian ZQ, De Young NJ, Liu JF, Mayne GC, Ruszkiewicz AR, Watson DI, Jamieson GG (2005) Metallothionien 3 expression is frequently down-regulated in oesophageal squamous cell carcinoma by DNA methylation. Mol Cancer 4:42

Smith DJ, Jaggi M, Zhang W, Galich A, Du C, Sterrett SP, Smith LM, Balaji KC (2006) Metallothioneins and resistance to cisplatin and radiation in prostate cancer. Urology 67 (6):1341–1347

Snyers L, Content J (1994) Induction of metallothionein and stomatin by interleukin-6 and glucocorticoids in a human amniotic cell line. Eur J Biochem 223(2):411–418

Sogawa CA, Asanuma M, Sogawa N, Miyazaki I, Nakanishi T, Furuta H, Ogawa N (2001) Localization, regulation, and function of metallothionein-III/growth inhibitory factor in the brain. Acta Med Okayama 55(1):1–9

Somji S, Sens MA, Lamm DL, Garrett SH, Sens DA (2001) Metallothionein isoform 1 and 2 gene expression in the human bladder: evidence for upregulation of MT-1X mRNA in bladder cancer. Cancer Detect Prev 25(1):62–75

Somji S, Garrett SH, Zhou XD, Zheng Y, Sens DA, Sens MA (2010) Absence of Metallothionein 3 Expression in Breast Cancer is a Rare, But Favorable Marker of Outcome that is Under Epigenetic Control. Toxicol Environ Chem 92(9):1673–1695

Somji S, Garrett SH, Toni C, Zhou XD, Zheng Y, Ajjimaporn A, Sens MA, Sens DA (2011) Differences in the epigenetic regulation of MT-3 gene expression between parental and Cd+2 or As+3 transformed human urothelial cells. Cancer Cell Int 11(1):2

Sone T, Koizumi S, Kimura M (1988) Cadmium-induced synthesis of metallothioneins in human lymphocytes and monocytes. Chem Biol Interact 66(1–2):61–70

Soo ET, Ng CT, Yip GW, Koo CY, Nga ME, Tan PH, Bay BH (2011) Differential expression of metallothionein in gastrointestinal stromal tumors and gastric carcinomas. Anat Rec (Hoboken) 294(2):267–272

Spiering R, van der Zee R, Wagenaar J, Kapetis D, Zolezzi F, van Eden W, Broere F (2012) Tolerogenic dendritic cells that inhibit autoimmune arthritis can be induced by a combination of carvacrol and thermal stress. PLoS One 7(9), e46336

Spiering R, Wagenaar-Hilbers J, Huijgen V, van der Zee R, van Kooten PJ, van Eden W, Broere F (2014) Membrane-bound metallothionein 1 of murine dendritic cells promotes the expansion of regulatory T cells in vitro. Toxicol Sci 138(1):69–75

Stasiolek M (2011) The role of selected immunoregulatory cell populations in autoimmune demyelination. Neuro Endocrinol Lett 32(1):25–33

Stasiolek M, Gavrilyuk V, Sharp A, Horvath P, Selmaj K, Feinstein DL (2000) Inhibitory and stimulatory effects of lactacystin on expression of nitric oxide synthase type 2 in brain glial cells. The role of Ikappa B-beta. J Biol Chem 275(32):24847–24856

Stasiolek M, Bayas A, Kruse N, Wieczarkowiecz A, Toyka KV, Gold R, Selmaj K (2006) Impaired maturation and altered regulatory function of plasmacytoid dendritic cells in multiple sclerosis. Brain 129(Pt 5):1293–1305

Stennard FA, Holloway AF, Hamilton J, West AK (1994) Characterisation of six additional human metallothionein genes. Biochim Biophys Acta 1218(3):357–365

Sugiura T, Kuroda E, Yamashita U (2004) Dysfunction of macrophages in metallothionein-knock out mice. J UOEH 26(2):193–205

Summer KH, Klein D, de Ruiter N, Abel J (1989) Metallothionein induction by nonsteroidal antiinflammatory drugs. Biol Trace Elem Res 21:165–169

Sun D, Ben-Nun A, Wekerle H (1988) Regulatory circuits in autoimmunity: recruitment of counter-regulatory CD8+ T cells by encephalitogenic CD4+ T line cells. Eur J Immunol 18 (12):1993–1999

Sunada F, Itabashi M, Ohkura H, Okumura T (2005) p53 negativity, CDC25B positivity, and metallothionein negativity are predictors of a response of esophageal squamous cell carcinoma to chemoradiotherapy. World J Gastroenterol 11(36):5696–5700

Sunardhi-Widyaputra S, van den Oord JJ, Van Houdt K, De Ley M, Van Damme B (1995) Identification of Metallothionein- and parathyroid hormone-related peptide (PTHrP)-positive cells in salivary gland tumours. Pathol Res Pract 191(11):1092–1098

Sundelin K, Jadner M, Norberg-Spaak L, Davidsson A, Hellquist HB (1997) Metallothionein and Fas (CD95) are expressed in squamous cell carcinoma of the tongue. Eur J Cancer 33 (11):1860–1864

Surowiak P, Dziegiel P, Matkowski R, Kornafel J, Wojnar A, Zabe M (2002) Immunocytochemical evaluation of metallothionein (MT) expression in myoepithelial cells of ductal mammary carcinoma and its relation to survival time: analysis of 7-year course of the disease. Folia Histochem Cytobiol 40(2):199–200

Surowiak P, Kaplenko I, Spaczynski M, Zabel M (2003) The expression of metallothionein (MT) and proliferation intensity in ovarian cancers treated with cisplatin and paclitaxel. Folia Morphol (Warsz) 62(4):493–495

Surowiak P, Matkowski R, Materna V, Gyorffy B, Wojnar A, Pudelko M, Dziegiel P, Kornafel J, Zabel M (2005) Elevated metallothionein (MT) expression in invasive ductal breast cancers predicts tamoxifen resistance. Histol Histopathol 20(4):1037–1044

Surowiak P, Materna V, Gyorffy B, Matkowski R, Wojnar A, Maciejczyk A, Paluchowski P, Dziegiel P, Pudelko M, Kornafel J, Dietel M, Kristiansen G, Zabel M, Lage H (2006) Multivariate analysis of oestrogen receptor alpha, pS2, metallothionein and CD24 expression in invasive breast cancers. Br J Cancer 95(3):339–346

Surowiak P, Materna V, Maciejczyk A, Pudelko M, Markwitz E, Spaczynski M, Dietel M, Zabel M, Lage H (2007) Nuclear metallothionein expression correlates with cisplatin resistance of ovarian cancer cells and poor clinical outcome. Virchows Arch 450(3):279–285

Suzuki T, Umeyama T, Ohma C, Yamanaka H, Suzuki K, Nakajima K, Kimura M (1991) Immunohistochemical study of metallothionein in normal and benign prostatic hyperplasia of human prostate. Prostate 19(1):35–42

Suzuki JS, Nishimura N, Zhang B, Nakatsuru Y, Kobayashi S, Satoh M, Tohyama C (2003) Metallothionein deficiency enhances skin carcinogenesis induced by 7,12-dimethylbenz[a] anthracene and 12-O-tetradecanoylphorbol-13-acetate in metallothionein-null mice. Carcinogenesis 24(6):1123–1132

Suzuki-Kurasaki M, Okabe M, Kurasaki M (1997) Copper-metallothionein in the kidney of macular mice: a model for Menkes disease. J Histochem Cytochem 45(11):1493–1501

Swierzcek S, Abuknesha RA, Chivers I, Baranovska I, Cunningham P, Price RG (2004) Enzyme-immunoassay for the determination of metallothionein in human urine: application to environmental monitoring. Biomarkers 9(4–5):331–340

Szelachowska J, Dziegiel P, Jelen-Krzeszewska J, Jelen M, Tarkowski R, Wlodarska I, Spytkowska B, Gisterek I, Matkowski R, Kornafel J (2008) Prognostic significance of nuclear and cytoplasmic expression of metallothioneins as related to proliferative activity in squamous cell carcinomas of oral cavity. Histol Histopathol 23(7):843–851

Szelachowska J, Dziegiel P, Jelen-Krzeszewska J, Jelen M, Tarkowski R, Spytkowska B, Matkowski R, Kornafel J (2009) Correlation of metallothionein expression with clinical progression of cancer in the oral cavity. Anticancer Res 29(2):589–595

Tai SK, Tan OJ, Chow VT, Jin R, Jones JL, Tan PH, Jayasurya A, Bay BH (2003) Differential expression of metallothionein 1 and 2 isoforms in breast cancer lines with different invasive potential: identification of a novel nonsilent metallothionein-1H mutant variant. Am J Pathol 163(5):2009–2019

Takaba K, Saeki K, Suzuki K, Wanibuchi H, Fukushima S (2000) Significant overexpression of metallothionein and cyclin D1 and apoptosis in the early process of rat urinary bladder carcinogenesis induced by treatment with N-butyl-N-(4-hydroxybutyl)nitrosamine or sodium L-ascorbate. Carcinogenesis 21(4):691–700

Takahashi S (2012) Molecular functions of metallothionein and its role in hematological malignancies. J Hematol Oncol 5:41

Takahashi Y, Ogra Y, Suzuki KT (2005) Nuclear trafficking of metallothionein requires oxidation of a cytosolic partner. J Cell Physiol 202(2):563–569

Takaishi M, Shimada A, Suzuki JS, Satoh M, Nagase H (2010) Involvement of metallothionein (MT) as a biological protective factor against carcinogenesis induced by benzo[a]pyrene (B[a] P). J Toxicol Sci 35(2):225–230

Takano H, Inoue K, Yanagisawa R, Sato M, Shimada A, Morita T, Sawada M, Nakamura K, Sanbongi C, Yoshikawa T (2004) Protective role of metallothionein in acute lung injury induced by bacterial endotoxin. Thorax 59(12):1057–1062

Tan Y, Sinniah R, Bay BH, Singh G (1999) Expression of metallothionein and nuclear size in discrimination of malignancy in mucinous ovarian tumors. Int J Gynecol Pathol 18(4):344–350

Tan OJ, Bay BH, Chow VT (2005) Differential expression of metallothionein isoforms in nasopharyngeal cancer and inhibition of cell growth by antisense down-regulation of metallothionein-2A. Oncol Rep 13(1):127–131

Tao X, Zheng JM, Xu AM, Chen XF, Zhang SH (2007) Downregulated expression of metallothionein and its clinicopathological significance in hepatocellular carcinoma. Hepatol Res 37(10):820–827

Tarapore P, Shu Y, Guo P, Ho SM (2011) Application of phi29 motor pRNA for targeted therapeutic delivery of siRNA silencing metallothionein-IIA and survivin in ovarian cancers. Mol Ther 19(2):386–394

Tekur S, Ho SM (2002) Ribozyme-mediated downregulation of human metallothionein II (a) induces apoptosis in human prostate and ovarian cancer cell lines. Mol Carcinog 33 (1):44–55

Tews DS, Nissen A, Külgen C, Gaumann AK (2000) Drug resistance-associated factors in primary and secondary glioblastomas and their precursor tumors. J Neurooncol 50(3):227–237

Theocharis S, Karkantaris C, Philipides T, Agapitos E, Gika A, Margeli A, Kittas C, Koutselinis A (2002) Expression of metallothionein in lung carcinoma: correlation with histological type and grade. Histopathology 40(2):143–151

Thirumoorthy N, Shyam Sunder A, Manisenthil Kumar K, Senthil Kumar M, Ganesh G, Chatterjee M (2011) A review of metallothionein isoforms and their role in pathophysiology. World J Surg Oncol 9:54

Thornalley PJ, Vasak M (1985) Possible role for metallothionein in protection against radiation-induced oxidative stress. Kinetics and mechanism of its reaction with superoxide and hydroxyl radicals. Biochim Biophys Acta 827(1):36–44

Tian ZQ, Xu YZ, Zhang YF, Ma GF, He M, Wang GY (2013) Effects of metallothionein-3 and metallothionein-1E gene transfection on proliferation, cell cycle, and apoptosis of esophageal cancer cells. Genet Mol Res 12(4):4595–4603

Tio L, Villarreal L, Atrian S, Capdevila M (2004) Functional differentiation in the mammalian metallothionein gene family: metal binding features of mouse MT4 and comparison with its paralog MT1. J Biol Chem 279(23):24403–24413

Tio L, Villarreal L, Atrian S, Capdevila M (2006) The Zn- and Cd-clusters of recombinant mammalian MT1 and MT4 metallothionein domains include sulfide ligands. Exp Biol Med (Maywood) 231(9):1522–1527

Tokar EJ, Diwan BA, Waalkes MP (2010) Early life inorganic lead exposure induces testicular teratoma and renal and urinary bladder preneoplasia in adult metallothionein-knockout mice but not in wild type mice. Toxicology 276(1):5–10

Tomita T (2000) Metallothionein in pancreatic endocrine neoplasms. Mod Pathol 13(4):389–395

Tomita T (2002) New markers for pancreatic islets and islet cell tumors. Pathol Int 52(7):425–432

Tomita T, Matsubara O (2000) Immunocytochemical localization of metallothionein in human pancreatic islets. Pancreas 20(1):21–24

Torlakovic EE, Bilalovic N, Golouh R, Zidar A, Angel S (2006) Prognostic significance of PU.1 in follicular lymphoma. J Pathol 209(3):352–359

Tran CD, Huynh H, van den Berg M, van der Pas M, Campbell MA, Philcox JC, Coyle P, Rofe AM, Butler RN (2003) Helicobacter-induced gastritis in mice not expressing metallothionein-I and II. Helicobacter 8(5):533–541

Tran CD, Campbell MA, Kolev Y, Chamberlain S, Huynh HQ, Butler RN (2005) Short-term zinc supplementation attenuates Helicobacter felis-induced gastritis in the mouse. J Infect 50 (5):417–424

Tran CD, Sundar S, Howarth GS (2009) Dietary zinc supplementation and methotrexate-induced small intestinal mucositis in metallothionein-knockout and wild-type mice. Cancer Biol Ther 8 (17):1662–1667

Trayhurn P, Duncan JS, Wood AM, Beattie JH (2000) Regulation of metallothionein gene expression and secretion in rat adipocytes differentiated from preadipocytes in primary culture. Horm Metab Res 32(11–12):542–547

Trivedy CR, Craig G, Warnakulasuriya S (2002) The oral health consequences of chewing areca nut. Addict Biol 7(1):115–125

Tsangaris GT, Tzortzatou-Stathopoulou F (1998) Metallothionein expression prevents apoptosis: a study with antisense phosphorothioate oligodeoxynucleotides in a human T cell line. Anticancer Res 18(4A):2423–2433

Tsangaris GT, Vamvoukakis J, Politis I, Kattamis AC, Tzortzatou-Stathopoulou F (2000) Metallothionein expression prevents apoptosis. II: Evaluation of the role of metallothionein expression on the chemotherapy-induced apoptosis during the treatment of acute leukemia. Anticancer Res 20(6B):4407–4411

Tse KY, Liu VW, Chan DW, Chiu PM, Tam KF, Chan KK, Liao XY, Cheung AN, Ngan HY (2009) Epigenetic alteration of the metallothionein 1E gene in human endometrial carcinomas. Tumour Biol 30(2):93–99

Tsuji T, Naito Y, Takagi T, Kugai M, Yoriki H, Horie R, Fukui A, Mizushima K, Hirai Y, Katada K, Kamada K, Uchiyama K, Handa O, Konishi H, Yagi N, Ichikawa H, Yanagisawa R, Suzuki JS, Takano H, Satoh M, Yoshikawa T (2013) Role of metallothionein in murine experimental colitis. Int J Mol Med 31(5):1037–1046

Tsujikawa K, Imai T, Kakutani M, Kayamori Y, Mimura T, Otaki N, Kimura M, Fukuyama R, Shimizu N (1991) Localization of metallothionein in nuclei of growing primary cultured adult rat hepatocytes. FEBS Lett 283(2):239–242

Tuccari G, Giuffre G, Arena F, Barresi G (2000) Immunohistochemical detection of metallothionein in carcinomatous and normal human gastric mucosa. Histol Histopathol 15 (4):1035–1041

Tüzel E, Kirkali Z, Yörükoglu K, Mungan MU, Sade M (2001) Metallothionein expression in renal cell carcinoma: subcellular localization and prognostic significance. J Urol 165(5):1710–1713

Tzankov A, Dirnhofer S (2006) Pathobiology of classical Hodgkin lymphoma. Pathobiology 73 (3):107–125

Uchida Y (1994) Growth-inhibitory factor, metallothionein-like protein, and neurodegenerative diseases. Biol Signals 3(4):211–215

Uchida Y, Ihara Y (1995) The N-terminal portion of growth inhibitory factor is sufficient for biological activity. J Biol Chem 270(7):3365–3369

Uchida Y, Tomonaga M (1989) Neurotrophic action of Alzheimer's disease brain extract is due to the loss of inhibitory factors for survival and neurite formation of cerebral cortical neurons. Brain Res 481(1):190–193

Uchida Y, Takio K, Titani K, Ihara Y, Tomonaga M (1991) The growth inhibitory factor that is deficient in the Alzheimer's disease brain is a 68 amino acid metallothionein-like protein. Neuron 7(2):337–347

Usvasalo A, Elonen E, Saarinen-Pihkala UM, Räty R, Harila-Saari A, Koistinen P, Savolainen ER, Knuutila S, Hollmén J (2010) Prognostic classification of patients with acute lymphoblastic leukemia by using gene copy number profiles identified from array-based comparative genomic hybridization data. Leuk Res 34(11):1476–1482

Valko M, Rhodes CJ, Moncol J, Izakovic M, Mazur M (2006) Free radicals, metals and antioxidants in oxidative stress-induced cancer. Chem Biol Interact 160(1):1–40

Vandeghinste N, Proost P, De Ley M (2000) Metallothionein isoform gene expression in zinc-treated human peripheral blood lymphocytes. Cell Mol Biol (Noisy-le-grand) 46(2):419–433

Varela M, Sala M, Llovet JM, Bruix J (2003) Treatment of hepatocellular carcinoma: is there an optimal strategy? Cancer Treat Rev 29(2):99–104

Vasak M (2005) Advances in metallothionein structure and functions. J Trace Elem Med Biol 19 (1):13–17

Vasak M, Meloni G (2011) Chemistry and biology of mammalian metallothioneins. J Biol Inorg Chem 16(7):1067–1078

Vazquez-Ramirez FJ, Gonzalez-Campora JJ, Hevia-Alvarez E, Fernandez-Santos JM, Rios-Martin JJ, Otal-Salaverri C, Gonzalez-Campora R (2000) P-glycoprotein, metallothionein and NM23 protein expressions in breast carcinoma. Pathol Res Pract 196(8):553–559

Vivanco I, Sawyers CL (2002) The phosphatidylinositol 3-Kinase AKT pathway in human cancer. Nat Rev Cancer 2(7):489–501

Waalkes MP, Harvey MJ, Klaassen CD (1984) Relative in vitro affinity of hepatic metallothionein for metals. Toxicol Lett 20(1):33–39

Waalkes MP, Liu J, Goyer RA, Diwan BA (2004) Metallothionein-I/II double knockout mice are hypersensitive to lead-induced kidney carcinogenesis: role of inclusion body formation. Cancer Res 64(21):7766–7772

Waalkes MP, Liu J, Kasprzak KS, Diwan BA (2005) Metallothionein-I/II double knockout mice are no more sensitive to the carcinogenic effects of nickel subsulfide than wild-type mice. Int J Toxicol 24(4):215–220

Waalkes MP, Liu J, Kasprzak KS, Diwan BA (2006) Hypersusceptibility to cisplatin carcinogenicity in metallothionein-I/II double knockout mice: production of hepatocellular carcinoma at clinically relevant doses. Int J Cancer 119(1):28–32

Waelput W, Broekaert D, Vandekerckhove J, Brouckaert P, Tavernier J, Libert C (2001) A mediator role for metallothionein in tumor necrosis factor-induced lethal shock. J Exp Med 194(11):1617–1624

Walentowicz-Sadlecka M, Koper A, Krystyna G, Koper K, Basta P, Mach P, Skret-Magierlo J, Dutsch-Wicherek M, Sikora J, Grabiec M, Kazmierczak W, Wicherek L (2013) The analysis of metallothionein immunoreactivity in stromal fibroblasts and macrophages in cases of uterine cervical carcinoma with respect to both the local and distant spread of the disease. Am J Reprod Immunol 70(3):253–261

Wan YY (2014) GATA3: a master of many trades in immune regulation. Trends Immunol 35 (6):233–242

Wang H, Zhang Q, Cai B, Li H, Sze KH, Huang ZX, Wu HM, Sun H (2006) Solution structure and dynamics of human metallothionein-3 (MT-3). FEBS Lett 580(3):795–800

Wei D, Fabris D, Fenselau C (1999) Covalent sequestration of phosphoramide mustard by metallothionein--an in vitro study. Drug Metab Dispos 27(7):786–791

Wei H, Desouki MM, Lin S, Xiao D, Franklin RB, Feng P (2008) Differential expression of metallothioneins (MTs) 1, 2, and 3 in response to zinc treatment in human prostate normal and malignant cells and tissues. Mol Cancer 7:7

Weinlich G, Zelger B (2007) Metallothionein overexpression, a highly significant prognostic factor in thin melanoma. Histopathology 51(2):280–283

Weinlich G, Bitterlich W, Mayr V, Fritsch PO, Zelger B (2003) Metallothionein-overexpression as a prognostic factor for progression and survival in melanoma. A prospective study on 520 patients. Br J Dermatol 149(3):535–541

Weinlich G, Eisendle K, Hassler E, Baltaci M, Fritsch PO, Zelger B (2006) Metallothionein - overexpression as a highly significant prognostic factor in melanoma: a prospective study on 1270 patients. Br J Cancer 94(6):835–841

Weinlich G, Topar G, Eisendle K, Fritsch PO, Zelger B (2007) Comparison of metallothionein-overexpression with sentinel lymph node biopsy as prognostic factors in melanoma. J Eur Acad Dermatol Venereol 21(5):669–677

Werynska B, Pula B, Muszczynska-Bernhard B, Piotrowska A, Jethon A, Podhorska-Okolow M, Dziegiel P, Jankowska R (2011) Correlation between expression of metallothionein and expression of Ki-67 and MCM-2 proliferation markers in non-small cell lung cancer. Anticancer Res 31(9):2833–2839

Werynska B, Pula B, Muszczynska-Bernhard B, Gomulkiewicz A, Jethon A, Podhorska-Okolow-M, Jankowska R, Dziegiel P (2013a) Expression of metallothionein-III in patients with non-small cell lung cancer. Anticancer Res 33(3):965–974

Werynska B, Pula B, Muszczynska-Bernhard B, Gomulkiewicz A, Piotrowska A, Prus R, Podhorska-Okolow M, Jankowska R, Dziegiel P (2013b) Metallothionein 1F and 2A overexpression predicts poor outcome of non-small cell lung cancer patients. Exp Mol Pathol 94(1):301–308

West AK, Hidalgo J, Eddins D, Levin ED, Aschner M (2008) Metallothionein in the central nervous system: Roles in protection, regeneration and cognition. Neurotoxicology 29 (3):489–503

West AK, Leung JY, Chung RS (2011) Neuroprotection and regeneration by extracellular metallothionein via lipoprotein-receptor-related proteins. J Biol Inorg Chem 16(7):1115–1122

Wicherek L, Popiela TJ, Galazka K, Dutsch-Wicherek M, Opławski M, Basta A, Klimek M (2005) Metallothionein and RCAS1 expression in comparison to immunological cells activity in endometriosis, endometrial adenocarcinoma and endometrium according to menstrual cycle changes. Gynecol Oncol 99(3):622–630

Widyarini S, Allanson M, Gallagher NL, Pedley J, Boyle GM, Parsons PG, Whiteman DC, Walker C, Reeve VE (2006) Isoflavonoid photoprotection in mouse and human skin is dependent on metallothionein. J Invest Dermatol 126(1):198–204

Wilhelmsen TW, Olsvik PA, Hansen BH, Andersen RA (2002) Evidence for oligomerization of metallothioneins in their functional state. J Chromatogr A 979(1–2):249–254

Wojnar A, Kobierzycki C, Krolicka A, Pula B, Podhorska-Okolow M, Dziegiel P (2010) Correlation of Ki-67 and MCM-2 proliferative marker expression with grade of histological malignancy (G) in ductal breast cancers. Folia Histochem Cytobiol 48(3):442–446

Wojnar A, Pula B, Piotrowska A, Jethon A, Kujawa K, Kobierzycki C, Rys J, Podhorska-Okolow M, Dziegiel P (2011) Correlation of intensity of MT-I/II expression with Ki-67 and MCM-2 proteins in invasive ductal breast carcinoma. Anticancer Res 31(9):3027–3033

Wolf C, Strenziok R, Kyriakopoulos A (2009) Elevated metallothionein-bound cadmium concentrations in urine from bladder carcinoma patients, investigated by size exclusion chromatography-inductively coupled plasma mass spectrometry. Anal Chim Acta 631 (2):218–222

Wong HR, Shanley TP, Sakthivel B, Cvijanovich N, Lin R, Allen GL, Thomas NJ, Doctor A, Kalyanaraman M, Tofil NM, Penfil S, Monaco M, Tagavilla MA, Odoms K, Dunsmore K, Barnes M, Aronow BJ, Genomics of Pediatric SSSI (2007) Genome-level expression profiles in pediatric septic shock indicate a role for altered zinc homeostasis in poor outcome. Physiol Genomics 30(2):146–155

Woolston CM, Deen S, Al-Attar A, Shehata M, Chan SY, Martin SG (2010) Redox protein expression predicts progression-free and overall survival in ovarian cancer patients treated with platinum-based chemotherapy. Free Radic Biol Med 49(8):1263–1272

Wróbel T, Mazur G, Dziegiel P, Surowiak P, Kuliczkowski K, Zabel M (2004) Expression of metallothionein (MT) and gluthatione s-transferase pi (SGTP) in the bone marrow of patients with myeloproliferative disorders (MPD). Folia Morphol (Warsz) 63(1):129–131

Wu Y, Siadaty MS, Berens ME, Hampton GM, Theodorescu D (2008) Overlapping gene expression profiles of cell migration and tumor invasion in human bladder cancer identify metallothionein 1E and nicotinamide N-methyltransferase as novel regulators of cell migration. Oncogene 27(52):6679–6689

Wülfing C, van Ahlen H, Eltze E, Piechota H, Hertle L, Schmid KW (2007) Metallothionein in bladder cancer: correlation of overexpression with poor outcome after chemotherapy. World J Urol 25(2):199–205

Xia N, Liu L, Yi X, Wang J (2009) Studies of interaction of tumor suppressor p53 with apo-MT using surface plasmon resonance. Anal Bioanal Chem 395(8):2569–2575

Yamada M, Hayashi S, Hozumi I, Inuzuka T, Tsuji S, Takahashi H (1996) Subcellular localization of growth inhibitory factor in rat brain: light and electron microscopic immunohistochemical studies. Brain Res 735(2):257–264

Yamamoto M, Tsujinaka T, Shiozaki H, Doki Y, Tamura S, Inoue M, Hirao M, Monden M (1999) Metallothionein expression correlates with the pathological response of patients with esophageal cancer undergoing preoperative chemoradiation therapy. Oncology 56(4):332–337

Yamasaki Y, Smith C, Weisz D, van Huizen I, Xuan J, Moussa M, Stitt L, Hideki S, Cherian MG, Izawa JI (2006) Metallothionein expression as prognostic factor for transitional cell carcinoma of bladder. Urology 67(3):530–535

Yamasaki M, Nomura T, Sato F, Mimata H (2007) Metallothionein is up-regulated under hypoxia and promotes the survival of human prostate cancer cells. Oncol Rep 18(5):1145–1153

Yan DW, Fan JW, Yu ZH, Li MX, Wen YG, Li DW, Zhou CZ, Wang XL, Wang Q, Tang HM, Peng ZH (2012) Downregulation of metallothionein 1F, a putative oncosuppressor, by loss of heterozygosity in colon cancer tissue. Biochim Biophys Acta 1822(6):918–926

Yanagie H, Hisa T, Ogata A, Miyazaki A, Nonaka Y, Nishihira T, Osada I, Sairennji T, Sugiyama H, Furuya Y, Kidani Y, Takamoto S, Takahashi H, Eriguchi M (2009) Improvement of sensitivity to platinum compound with siRNA knockdown of upregulated genes in platinum complex-resistant ovarian cancer cells in vitro. Biomed Pharmacother 63(8):553–560

Yang Y, Maret W, Vallee BL (2001) Differential fluorescence labeling of cysteinyl clusters uncovers high tissue levels of thionein. Proc Natl Acad Sci USA 98(10):5556–5559

Yang F, Zhou M, He Z, Liu X, Sun L, Sun Y, Chen Z (2007) High-yield expression in Escherichia coli of soluble human MT2A with native functions. Protein Expr Purif 53(1):186–194

Yap X, Tan HY, Huang J, Lai Y, Yip GW, Tan PH, Bay BH (2009) Over-expression of metallothionein predicts chemoresistance in breast cancer. J Pathol 217(4):563–570

Yin X, Knecht DA, Lynes MA (2005) Metallothionein mediates leukocyte chemotaxis. BMC Immunol 6:21

Youn J, Lynes MA (1999) Metallothionein-induced suppression of cytotoxic T lymphocyte function: an important immunoregulatory control. Toxicol Sci 52(2):199–208

Youn J, Borghesi LA, Olson EA, Lynes MA (1995) Immunomodulatory activities of extracellular metallothionein. II Effects on macrophage functions J Toxicol Environ Health 45(4):397–413

Youn J, Hwang SH, Ryoo ZY, Lynes MA, Paik DJ, Chung HS, Kim HY (2002) Metallothionein suppresses collagen-induced arthritis via induction of TGF-beta and down-regulation of proinflammatory mediators. Clin Exp Immunol 129(2):232–239

Yu X, Wu Z, Fenselau C (1995) Covalent sequestration of melphalan by metallothionein and selective alkylation of cysteines. Biochemistry 34(10):3377–3385

Yu WH, Lukiw WJ, Bergeron C, Niznik HB, Fraser PE (2001) Metallothionein III is reduced in Alzheimer's disease. Brain Res 894(1):37–45

Yu J, Fujishiro H, Miyataka H, Oyama TM, Hasegawa T, Seko Y, Miura N, Himeno S (2009) Dichotomous effects of lead acetate on the expression of metallothionein in the liver and kidney of mice. Biol Pharm Bull 32(6):1037–1042

Yuguchi T, Kohmura E, Yamada K, Sakaki T, Yamashita T, Otsuki H, Kataoka K, Tsuji S, Hayakawa T (1995) Expression of growth inhibitory factor mRNA following cortical injury in rat. J Neurotrauma 12(3):299–306

Yuguchi T, Kohmura E, Sakaki T, Nonaka M, Yamada K, Yamashita T, Kishiguchi T, Sakaguchi T, Hayakawa T (1997) Expression of growth inhibitory factor mRNA after focal ischemia in rat brain. J Cereb Blood Flow Metab 17(7):745–752

Yurkow EJ, DeCoste CJ (1999) Effects of cadmium on metallothionein levels in human peripheral blood leukocytes: a comparison with zinc. J Toxicol Environ Health A 58(5):313–327

Yurkow EJ, Makhijani PR (1998) Flow cytometric determination of metallothionein levels in human peripheral blood lymphocytes: utility in environmental exposure assessment. J Toxicol Environ Health A 54(6):445–457

Zagorianakou N, Stefanou D, Makrydimas G, Zagorianakou P, Briasoulis E, Karavasilis V, Pavlidis N, Agnantis NJ (2006) Clinicopathological study of metallothionein immunohisto-chemical expression, in benign, borderline and malignant ovarian epithelial tumors. Histol Histopathol 21(4):341–347

Zaia J, Jiang L, Han MS, Tabb JR, Wu Z, Fabris D, Fenselau C (1996) A binding site for chlorambucil on metallothionein. Biochemistry 35(9):2830–2835

Zalups RK, Ahmad S (2003) Molecular handling of cadmium in transporting epithelia. Toxicol Appl Pharmacol 186(3):163–188

Zamirska A, Matusiak L, Dziegiel P, Szybejko-Machaj G, Szepietowski JC (2012) Expression of metallothioneins in cutaneous squamous cell carcinoma and actinic keratosis. Pathol Oncol Res 18(4):849–855

Zangger K, Shen G, Oz G, Otvos JD, Armitage IM (2001) Oxidative dimerization in metallothionein is a result of intermolecular disulphide bonds between cysteines in the alpha-domain. Biochem J 359(Pt 2):353–360

Zbinden S, Wang J, Adenika R, Schmidt M, Tilan JU, Najafi AH, Peng X, Lassance-Soares RM, Iantorno M, Morsli H, Gercenshtein L, Jang GJ, Epstein SE, Burnett MS (2010) Metallothionein enhances angiogenesis and arteriogenesis by modulating smooth muscle cell and macrophage function. Arterioscler Thromb Vasc Biol 30(3):477–482

Zeng J, Heuchel R, Schaffner W, Kagi JH (1991a) Thionein (apometallothionein) can modulate DNA binding and transcription activation by zinc finger containing factor Sp1. FEBS Lett 279 (2):310–312

Zeng J, Vallee BL, Kagi JH (1991b) Zinc transfer from transcription factor IIIA fingers to thionein clusters. Proc Natl Acad Sci USA 88(22):9984–9988

Zhang XH, Takenaka I (1998) Incidence of apoptosis and metallothionein expression in renal cell carcinoma. Br J Urol 81(1):9–13

Zhang B, Satoh M, Nishimura N, Suzuki JS, Sone H, Aoki Y, Tohyama C (1998) Metallothionein deficiency promotes mouse skin carcinogenesis induced by 7,12-dimethylbenz[a]anthracene. Cancer Res 58(18):4044–4046

Zhang R, Zhang H, Wei H, Luo X (2000) Expression of metallothionein in invasive ductal breast cancer in relation to prognosis. J Environ Pathol Toxicol Oncol 19(1–2):95–97

Zhang Y, Ma CJ, Wang JM, Ji XJ, Wu XY, Moorman JP, Yao ZQ (2012) Tim-3 regulates pro- and anti-inflammatory cytokine expression in human CD14+ monocytes. J Leukoc Biol 91 (2):189–196

Zheng H, Liu J, Choo KH, Michalska AE, Klaassen CD (1996a) Metallothionein-I and -II knock-out mice are sensitive to cadmium-induced liver mRNA expression of c-jun and p53. Toxicol Appl Pharmacol 136(2):229–235

Zheng H, Liu J, Liu Y, Klaassen CD (1996b) Hepatocytes from metallothionein-I and II knock-out mice are sensitive to cadmium- and tert-butylhydroperoxide-induced cytotoxicity. Toxicol Lett 87(2–3):139–145

Zheng Q, Yang WM, Yu WH, Cai B, Teng XC, Xie Y, Sun HZ, Zhang MJ, Huang ZX (2003) The effect of the EAAEAE insert on the property of human metallothionein-3. Protein Eng 16 (12):865–870

Zhou XD, Sens DA, Sens MA, Namburi VB, Singh RK, Garrett SH, Somji S (2006a) Metallothionein-1 and -2 expression in cadmium- or arsenic-derived human malignant urothelial cells and tumor heterotransplants and as a prognostic indicator in human bladder cancer. Toxicol Sci 91(2):467–475

Zhou XD, Sens MA, Garrett SH, Somji S, Park S, Gurel V, Sens DA (2006b) Enhanced expression of metallothionein isoform 3 protein in tumor heterotransplants derived from As+3- and Cd+2-transformed human urothelial cells. Toxicol Sci 93(2):322–330

Ziegler-Heitbrock L (2007) The CD14+ CD16+ blood monocytes: their role in infection and inflammation. J Leukoc Biol 81(3):584–592

Zilliox MJ, Parmigiani G, Griffin DE (2006) Gene expression patterns in dendritic cells infected with measles virus compared with other pathogens. Proc Natl Acad Sci USA 103 (9):3363–3368

Znidaric MT, Pucer A, Fatur T, Filipic M, Scancar J, Falnoga I (2007) Metal binding of metallothioneins in human astrocytomas (U87 MG, IPDDC-2A). Biometals 20(5):781–792

Printed in the United States
By Bookmasters